Television Operations Handbook

ROBERT S. ORINGEL

FOCAL PRESS
Boston • London

FOCAL PRESS
is an imprint of Butterworths, which has principal offices
in Boston and London.

Library of Congress Cataloging in Publication Data

Oringel, Robert S.
   Television operations handbook.

   Includes index.
   1. Television.   I. Title.
TK6630.075 1984   621.388   84–10140
ISBN 0–240–51734–2

Butterworth Publishers
80 Montvale Ave.
Stoneham, MA 02180

10 9 8 7 6 5 4 3 2

Printed in the United States of America

This book is dedicated to those who listen to radio and
watch television; to those who were fortunate enough
to spend their working lives creating the broadcast arts
and sciences; and to those yet to come, who will learn
and love broadcasting.

Contents

Contents   vii

Preface and Acknowledgments

This text should be looked upon as anything but a scholarly work. It is, rather, a compilation in one source of a variety of information that has been available—however scattered—for many years, in the various texts, pamphlets, brochures, and spec. sheets which describe the many phases of technical/operational television.

This author decided, as a member of a local government advisory task force, that in order to train large numbers of people for cable television local access programming, the material should be available in one place, and at a particular level of instruction (see column 1, page 1).

We use the male gender in most of this text, when we describe a camera *man*, for instance. We apologize and hasten to state that no male chauvinism is intended. It is just so much easier to let the words flow, than the use of "his/hers", or a neuter description like cameraperson. We readily acknowledge the existence of most able women technicians and production people in this field.

The author has been assisted in this undertaking by the extreme kindness of many video and audio equipment manufacturers and individual members of their staffs. These, named below, have supplied and permitted the use of verbatim description, direct quotation and paraphrasing from their very excellent sales promotional material and instruction booklets, as well as providing photos and diagrams of their equipment for inclusion in this text.

The author is grateful as well to his friends and colleagues in the field, who have read the manuscript and provided guidance.

It is therefore with a deep and most humble bow that my thanks go to the following:

Bernard Singer, friend and former USIA colleague, retired Foreign Service Officer, television specialist, who read and reread and made corrections to the manuscript until I got it right.

Sue Buske, the executive director of the National Federation of Local Cable Programmers, a critical reader of the manuscript.

Herbert C. Pearson, manager of the Division of Television and Audio-Visual

Services at Georgetown University, Washington, D.C., a critical reader of the manuscript.

Malarkey-Taylor Associates, Inc., Telecommunications Consultants, and its Vice-President for Regulatory Affairs, Gary S. Hurwitz, who provided permission to quote from its excellent report to the city of Bowie, MD

The Manufacturers

American Data; Amperex; Ampex; Angenieux; Atlas Stands; Colortran; Comrex; Crosspoint Latch; Electro-Voice; Frezzolini; Hitachi Denshi America Ltd., and Mary Jo Lyle of its Lanham MD office; Ikegami; Industrial Sciences Inc. (ISI); JVC; Knox; Leitch; Listec; 3 M Company; Micron; Microtime; Minolta; Neutrik; Panasonic; Quick-Set; RCA; Sennheiser; Shure Brothers; The Sony Corporation of America, its Lanham MD Sales Engineer David J. Doherty and Theresa Dabrowski, Production Coordinator of Park Ridge, NJ; Switchcraft; Teac; Tektronix Inc. and its Rockville, MD representative, James A. Capps; Walt Disney Productions; Ward-Beck Systems.

Finally, a huge measure of thanks to Mr. Isaac Creek, photographer/teacher, and his students at Eleanor Roosevelt High School, Prince Georges County, MD, for the photos depicting the placement of lights.

Robert S. Oringel

Chapter 1
Television: Two Reception Modes

BROADCAST TELEVISION

This book began with an idea that encompassed a very specific, narrow purpose—to teach and describe the operation of the television equipment which is used for the production of cable access channel television programs. Along the way, a strange thing happened. As the author began to ask some colleagues in the field to critique the manuscript, and later as a couple of editors at publishing houses read the manuscript, the same question was asked: Why deliberately limit the book to a *cable* television audience? In each case this was followed by the statement, "The techniques described in the book are equally valid for all television operation, not solely cable." Then the editors said, "A broader audience will make the book more viable."

The author knew that the comments were correct and was immediately convinced. So, if you have no specific interest in access cable or cable television but are interested in broadcast television operation only, go to the end of this chapter, where we discuss the television production crew.

CABLE TELEVISION

After World War II, as the new communication medium of broadcast television changed from a novelty that most people watched in furniture and radio repair store windows to a virtual necessity found in almost every U.S. home, a serious viewing problem arose for those people living in mountainous areas. Broadcast television existed, but its reception was blocked by nearby hills and mountains.

A very enterprising television sales and service store owner in Lansford, Pennsylvania, northwest of Allentown, built a receiving antenna on a tower atop a local hill. He offered to connect the television customers in the nearby valley to his "community" antenna, providing them with vastly improved reception for a small fee. With that simple stroke of business genius, community antenna television (CATV) came into being.

CATV soon became cable television, as we know it today, with many cable system operators offering both the improved reception of off-air broadcast television signals and ever-expanding additional services. Three concurrent technical developments allowed this to happen.

THE ROLE OF THE HELICAL SCAN VTR

First, the development of the relatively inexpensive, lightweight helical-scan video tape recorder (VTR) made taped news gathering possible—called electronic news gathering (ENG) and electronic field production (EFP)—by permitting the VTR to go out in the field with the video camera and crew.

THE INTRODUCTION OF SATELLITE NETWORKING

The second development was the commercial beginning of satellite transmission and reception of television programming. This momentous occurrence brought entertainment packaging and movie channel networking inexpensively to the cable operator. In turn, the cable operator could offer uncut movies and other popular entertainment to subscribers at a cost far lower than if the networking had had to travel intracoastally via coaxial cable.

400-MHZ TECHNOLOGY

The third development was the arrival of 400-megahertz cable transmission technology. This technology enabled cable operators to provide up to 102 channels of programming to subscribers, rather than the 30 plus channels previously available. Thus the amount of marketable product available to be sold increased, thereby increasing profitability to the cable operator.

CABLE ACCESS TV

Along with cable television entertainment came a community-perceived need for local communities to become involved in "narrow-casting"; that is, bringing television *information* in addition to entertainment into the home. At first the need was evoked by a requirement of the Federal Communications Commission (FCC) imposed on every new cable franchise. Later, the requirement by the FCC was dropped and was replaced by a public demand in most requests for proposals (RFPs) that channel(s) for public use be set aside by the cable operator as a part of the franchise contract.

This demand was manifested in a form of cable television known as *public access* television. The cable franchise granted by a community stipulates that the franchisee, the cable television operator, make available, usually through a nonprofit entity such as an access corporation or perhaps through a local library system, one or more cable access channels of the hundred plus channels available on the system, *including funding for programming*, to the community granting the franchise.

To the extent that it is successful—and that success depends mightily on two factors, outreach and technical professionalism—access involves a local community by using an all-volunteer community staff. This staff consists of production people, technicians, on-camera "talent," script writers, and all of the other people who are needed to put television programs on the air, on one or more access channels. Often the only paid professionals are the full-time top managerial people of the access corporation and the necessary clerical, training, and scheduling people.

Access programs vary in complexity

from the 5-minute or less sound-off sege-ments, in which individuals from the local community sit "on camera" in a studio and talk about various issues (within prescribed morality requirements), to local little league and high school ball games, to locally written and produced dramatic and variety programs. Usually plays and music written by known authors and composers are forbidden by the copyright holders to be performed on local access; this is because when the performances are videotaped, there is a monetary loss to the author or composer since the tape is replayed without compensation.

The incentive to community members to get involved in access is twofold. First, there is the fun of learning new skills and using them to *make* a program. Second, individuals can reach the community with their special interests: sports, gardening, stamp collecting, quilting, poetry reading, local gossip, strongly held personal opinions, or any one of a myriad of viewpoints and topics that find their way onto cable access channels.

From the point of view of the franchisee, who may be public-spirited enough to admire access but whose primary consideration must be business, it is usually preferable for the community to control access programming, rather than to have it included as an integral part of the operation, like the local-origination programming.

The cable operator derives no direct income from access. And if it fails as a viable community activity, usually as a result of lack of interest, the cable operator eventually gets the channel(s) originally devoted to access returned to him, to be converted to income-producing enterprises such as commercially leased channels. Note that providing access channels is very good public relations for a fran-

chisee; indeed, surveys indicate that some people subscribe to cable television primarily to view access television, which they see as a refreshing change from broadcast network programming. An indirect benefit to the cable operator, then, is the subscription of those people who buy the service because they use or view access and want cable in their homes.

THE CABLE FRANCHISING PROCESS

Cable television in a community begins with the franchising process, which is, unfortunately, long and drawn out. When the governing body of the community, often the city or county council, decides that its constituency wants cable television, it issues a request for proposals, which is generally printed in the cable television trade press. This RFP is coupled in most cases with a set of recommendations from either a professional consultant or a community cable TV committee of citizens, who provide the governing body with minimum requirements that should be demanded of the franchise applicants. Usually required is a biographic listing of all the principals in the applicant company, with the amount of voting stock owned by each, including how that stock was procured (free, for cash payment, or with borrowed money) and the identities of those principals living in the local community. This helps identify any "rent-a-citizen" who might have been given stock in the company, either free or at a reduced cost, to aid a prospective franchisee in the political aspects of the franchise process.

Also required from prospective franchisees is a list of the company's attested financial assets and liabilities. The financial factors in the choice of a franchisee are primary decision-making considerations. The technical and programming plans of the company, as specified in its documented reply to the RFP, may collapse if the company's financial base is weak or insufficient. Historically, in the cable television field, the lack of adequate financial resources and the mismanagement of those resources have been the major factors contributing to delays in plant construction and to poor operational performance by cable operators.

The elements to be evaluated in the financial consideration of an applicant company are its financial capability, attested to by banking or other financial institutions who guarantee its borrowing power; its sworn financial commitments to this specific community cable system; its financial obligations to other cable systems and other company projects; its financial projections and assumptions; its rate structure and the feasibility of rate stability; and finally a description of the viability of this system based on solid financial projections.

AN INSIGHT INTO CABLE TV ECONOMICS

To understand the financial aspects of cable, we must understand some of the economic factors that make up the complex monetary basis of cable television. The first factor in cable economics is *penetration*, defined as the percentage of homes passed by the cable system that *subscribe to its services*. It is most important in the success or failure of a system. if only a small percentage, say 10 percent or less, of the potential subscribers sign up for service, the system will assuredly fail, whereas a subscriber penetration rate of 38 to 40 percent most likely will indicate financial viability and penetration over 50 percent indicates assured success. The in-

dustry has found that the financial breakeven point is somewhere between 30 and 40 percent penetration. The likelihood of subscriber penetration is enhanced by extra-cost uncut TV movies and other satellite programming. There will be even greater penetration if the cable operator offers home security protection via the cable system and other interactive (two-way) services, such as the Columbus, Ohio, QUBE system. "Adult" movie channels together with appropriate lockout systems to prevent viewing by minors also enhance viewer penetration.

Density, or the number of households per cable mile, is the second major contributing factor to system viability. The system's density figure describes the number of potential subscribers among whom the capital costs and fixed operating expenses can be distributed. When the viability of a system is assessed, the interrelationship of density and penetration is important because tradeoffs between the two can affect viability. A low-density area may be viable if a very high penetration can be achieved, or medium penetration may be viable in a very high-density area. Careful analysis of the density and expected penetration will determine the level of service that the system can support, its probable cost, and its projected financial success or failure.

Income for a cable system is derived primarily from monthly charges for basic service, installation fees, and pay services such as movie channels, with additional revenue expected to be generated by the newer services such as home security, information services, data transfer services to subscribers with home computers, and advertising and leased business channels. Income for a system may be further enhanced by the addition of an *"institutional network,"* another interactive cable which carries information two ways between businesses, government bodies, and other entities that find it necessary to exchange other than telephone conversation on an ongoing daily basis.

The operating expenses of a cable television system, while comprising the usual business expenses of salaries, rent, utilities, and office supplies, also include copyright fees, pole rental fees to utility companies, fees to satellite program distributors based on a per capita subscriber rate, franchise fee payments to municipalities (which are from 3 to 5 percent of the subscriber revenue), and programming expenses. The operating expenses of a large system range from 50 to 60 percent of revenue, and the expenses of operating a small rural system range from 40 to 55 percent of revenue.

Capital expenditures include program distribution centers, called *headends* or *hubs*, the cable distribution system, and subscriber drops. Program distribution centers typically are composed of a receiving site with antenna tower(s) and satellite receiving "dishes" and a small building crammed with electronic equipment, offices, and studios. The studios contain costly lighting equipment, cameras and their control equipment, audio systems, video tape recorders, and videotape editing systems.

The cable distribution system has literally miles of high-priced coaxial cable, in trunk cables, feeder cables, and drop cables, all with associated distribution amplification equipment, running from the program distribution center to the subscribers' neighborhoods. The drops at subscriber homes include both converters and traps. *Converters* are interface devices between the cable system and the subscriber's television receiver. *Traps* are electronic devices that block reception of signals (pay TV channels) which the subscriber has not purchased. This distribution system is *the* major cost item in a cable television system, and that cost depends on how much of the system is installed aboveground, on poles leased from local utilities, and on how much of the systems must be dug (trenched and refilled) underground. An underground mile of system is much more expensive than an aerial mile of the same system.

Subscriber drop costs are determined primarily by the cost of the converter interfacing device. In an addressable two-way system, this can run to more than $200 each, and this cost is multiplied by the number of subscribers to which a converter must be supplied.

The total cost, therefore, of passing by a home with a cable television system ranges from $150 to more than $700; needless to say, these costs must be divided among the subscribers. That explains why the viability of the system is so dependent on the density, subscriber penetration, and percentage of underground plant required.

The third requirement of a prospective franchisee is an in-depth description of the proposed technical design of the entire physical plant. Because of its high cost the plant will change little, except for minor upgrading of technology, during the lifetime of the franchise, usually set at 15 years. Thus it is necessary to examine critically all technical aspects in order to ensure that the plant will meet or exceed the needs of the community, both at the time of system startup and during the entire term of the franchise.

Certain aspects of technical design must be considered:

- The quality of reception and the number of channels of off-air pickup broadcast television, satellite pickup, and interconnection facilities.
- The proposed bandwidth of the system in both the downstream (subscriber)

and the upstream (interactive) directions. Six megahertz of width is needed per downstream channel, and considerably less per upstream channel, plus the system's capability to add services, by using spare bandwidth, at a later date.

- The subscriber interfacing terminal(s) or converters proposed for use.

- The source or sources of emergency standby power proposed to replace the normal utility company power. This becomes most important when home security is a part of the cable system and a loss in utility power would leave the security system useless if it had no backup power sources.

- The proven technical performance of the system, the methods used in measuring that performance, the regularity of measurement, and the response to customer complaint about performance.

- The construction costs and practices to be employed in building the proposed system.

- The final requirement asked of a prospective franchisee is a description of the proposed cable programming and the other video services to be offered to the community. The applicant must detail the five customary categories of available program and service options and how those options will be delivered.

- The five categories comprise the retransmitted broadcast television signals, those required of a cable operator by both the FCC and others, including far-away stations; the satellite-delivered cable channels; the cable operator–supplied local-origination programming; the startup and continuing yearly monetary support offered to local access; and the enhanced and developmental interactive services.

Since all applicants for franchise undoubtedly will propose to provide, to one degree or another, all of the options mentioned, the evaluation should be based on the *tiering* structure of the options together with the rate structure provided for the options. Combined, they describe the cost to the subscriber of cable television and offer one major means of comparison among applicants.

Having discussed the structure of cable television at such length, we now focus down on access programming in cable and the ways in which it is technically accomplished.

To put access, or for that matter any, programming on the channels of a cable system requires the use of television equipment–operating skills, and teaching those skills is the real task of this text. We also wish to instill in our readers a sense of the aesthetics of television programming—the very real, discernible difference between a poorly produced or engineered program and one that is done with taste and respect for the sensibilities of the viewer. We should have little problem with that goal, since most readers have grown up watching well-produced, well-engineered broadcast television, even if the program content was often dreadful. It is our hope that what was good remained in memory; now with access, our readers have the power to respond and to create good program content as well.

To teach the use of television operating equipment, it is axiomatic that this book be read in the context of available equipment for hands-on operating. No one can learn to drive a car while sitting at home, regardless of the caliber of instruction. And the value of repeated experience cannot be overemphasized. Experience alone separates the access volunteer from the professional network technician.

Finally, before we get into the technicalities and techniques, let us look broadly at the people, or, more accurately, the jobs involved in placing an access cable TV program before a viewer.

**THE TELEVISION
PRODUCTION CREW**

The Producer. From the outset this person is responsible for the program. The producer combines the idea or the scripted material with the funding needed to do the job and hires the talent and a director. Then the entire job is put in the hands of the director, and the producer withdraws from the process unless he will be the director as well as the producer.

The Director. This person meets with the producer, accepts the directorial assignment, and takes command of the production. He explains directorial policies to the operating crew and to the talent. The director rehearses the talent, prepares the shooting script, and finally directs the production. During studio rehearsals and during the production, he provides standbys and cues for the video switcher, cameramen, the audio man and, through the floor manager, the cast. The primary quality of the director in dealing with talent and crew is that he is *the* authority figure.

The Technical Director. Usually the technical director (TD) is the technical supervisor of the crew. He operates the production switcher during the program and takes standbys and cues from the director. The technical director is attentive and ready to switch video inputs to the program.

The Audio Technician. This person operates the audio control console, mixes all of the audio inputs, and takes standbys and cues from the director during the rehearsal and the program. He often cues the microphone boom operators.

The Video Technician. The camera control units (CCUs) are operated by the video technician. He is responsible for video gain and color purity of the picture during the program.

The Floor Manager. This person must be the director's eyes, ears, and hands in the studio. He cues the talent and maintains order in the studio. Also the floor manager supervises striking (breaking down) the set after the program is concluded.

The Cameraman. This person follows a shot schedule exerpted from the director's shooting script and moves the camera to the assigned place for the next shot. He frames and focuses so that he is ready when the director cues his camera. The cameraman, above all, must be innovative in looking for shot variations that would enhance the program's pictures.

REVIEW QUESTIONS

1. What was the original purpose of CATV?

2. Why do we have community-access TV?

3. How does access differ from local-origination cable television?

4. How does the franchising process start?

5. What minimum requirements are demanded of a prospective franchisee?

6. State and describe the primary factors in cable economics that make a cable system feasible.

7. What jobs must be performed to bring a television program to the viewer?

Chapter 2

Light and Lenses

We begin our study of how to operate television production equipment by first learning how that equipment produces a television picture. A film camera picture is made by light entering a lens and striking a sensitized surface. A television picture is made in the very same way. So first we examine some of the properties of light and see how it is used in photography (video or film); then we examine lenses.

We discuss both light and lenses in this book only in enough depth to enable the newcomer to understand how the video camera works. There are many excellent photographic texts which discuss these subjects in great detail, and we do not repeat what is readily available to the inquisitive newcomer.

Light

DEFINITION OF LIGHT

Light may be defined as a form of electromagnetic radiation which travels in waves. Light travels at a speed of 186,000 miles per second in a vacuum and at practically the same speed in air. Its speed slows somewhat when it enters glass or water.

BEHAVIOR OF LIGHT

When waves of light hit an object, a number of things may happen. The light waves may be *reflected* (they may bounce off, as in a mirror), *refracted* (they bend and continue on their way, as when hitting glass or water), or they may be dispersed or absorbed by the object that they strike. (See Fig. 2–1.)

Our primary concern here is the *refraction of light*. Light waves are refracted, or bent, when they pass from one *transparent* substance to another transparent substance—from air to glass or from glass to water. During the transfer from one medium to another, the speed of the light waves and thus their wavelengths are

Figure 2–1 Lightwave behavior.

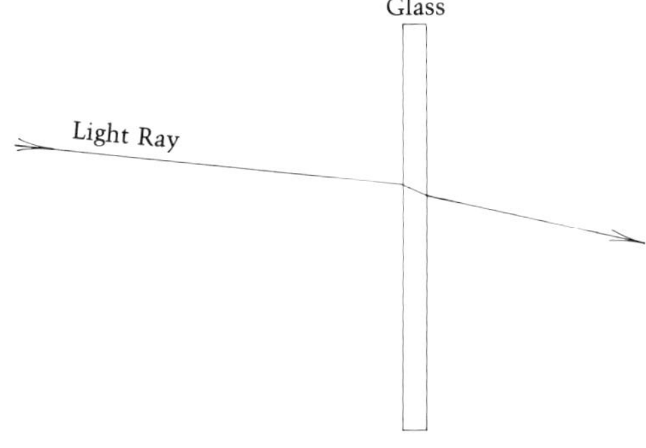

changed slightly, which causes the light to bend.

Refraction, then, is responsible for the behavior of lenses. It permits the light that is reflected from a wide panoramic scene to be bent into the small opening of a television camera; conversely, refraction allows the light from a powerful lamp source to be bent and dispersed over a wide area. When light is used on the *set*, or staging area, of a television program, we call it *lighting*, and the electric devices used in lighting are called *lamps, lighting fixtures*, or *instruments*.

Two broad categories of lighting are used in television: natural light (or sunlight) and artificial light (or electric incandescent light). They may be used together or separately.

Sunlight

The first category, sunlight, is often called *available light* even though it is not available indoors or at night. It can be controlled to some degree out of doors and made more available by the use of large panel reflectors, which aim reflected light from the sun at a subject or scene, when the need is to concentrate the light. Light, which activates the television pickup tube, may be diminished in intensity by the use of the *camera lens aperture* and by the neutral-density filter used to control the amount of light entering the lens. (See Fig. 2–2.) We discuss lenses and filters in greater depth later.

Incandescent Light

The second category is incandescent light. It is available to the video cameraman in several forms. These range in intensity

Figure 2–2 Sunlight reflector. *(Provided courtesy of Walt Disney Productions.)*

from the small battery-pack-operated, handheld (Fig. 2–3) or camera-top-mounted portable lamp; to portable light-weight, three-fixture kits; to massive banks of high-current studio lights called *"cyc's"* (from *cyclorama*), or *"cyc lights"* (because they are used to supply background or overall fill illumination to the cyclorama background curtain). (See Fig. 2–4.) Often these cyc lights are called *strip lighting*, and some are covered with col-

Figure 2–3 Handheld light. *(Provided courtesy of Colortran.)*

ored plastic *gelatin*, or "gel," to inject color (yellow gel for sunset, blue gel for early morning or early evening) into the scene.

Lighting Fixtures, Scrims, and Barn Doors

The most prevalent type of studio lamp is the tungsten halogen incandescent lamp. It is available in many wattages, and the individual lamp's light intensity can be controlled by the use of *dimmers*, or *scrims*. Dimmers are used only for cyclorama lighting and accent or background set lighting, where the color temperature of the light is not critical, since the use of dimmers will change the color temperature of the light. Scrims are diffusers, placed in front of the lamp, which soften and decrease the intensity of the light but maintain its color temperature at a steady 3200 to 3400 degrees Kelvin. (See Fig. 2–5.) This is the standard color temperature range of light to be maintained for accurate skin tones on color television. The color temperature of sunlight is approximately twice that of studio light, or around 6400 degrees Kelvin. A neutral-density filter is used to decrease the intensity, and color filters plus a neutral-density filter are used to decrease color temperature to the 3200 degrees Kelvin, at which the television camera is designed to operate best.

The concentration of light from lamps is controlled by mounting the lamps in shaped reflectors behind the light source; by placing concentrating lenses, called *Fresnel lenses*, in front of the light source; and, in lower powered and portable lamps, by using a control wheel or joystick at the back of the fixture which moves the lamp in and out as the reflector changes the focus of the light beam. Reflector shapes vary from the *scoop*, which is open at the front and bowl-shaped and can be either deep or shallow, to closed fixtures with Fresnel lenses in front of lamps of 1 to 5 kilowatts (KW). (See Figs. 2–6 and 2–7.)

Thus lighting fixtures fall into two categories: focused light, or nonfocused, diffused fill light. In focused light, fixtures

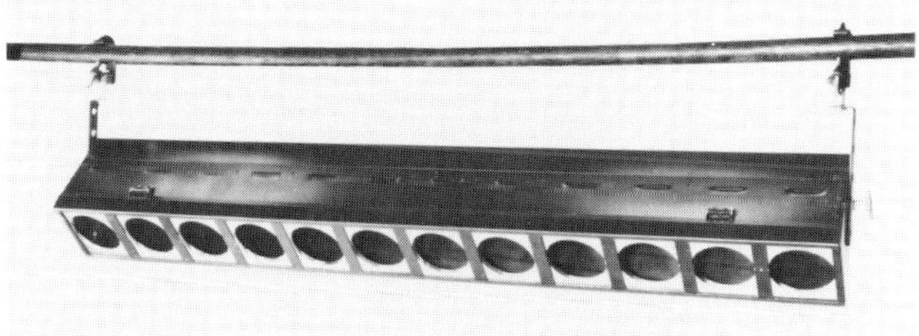

Figure 2–4 Cyc light banks. *(Provided courtesy of Colortran.)*

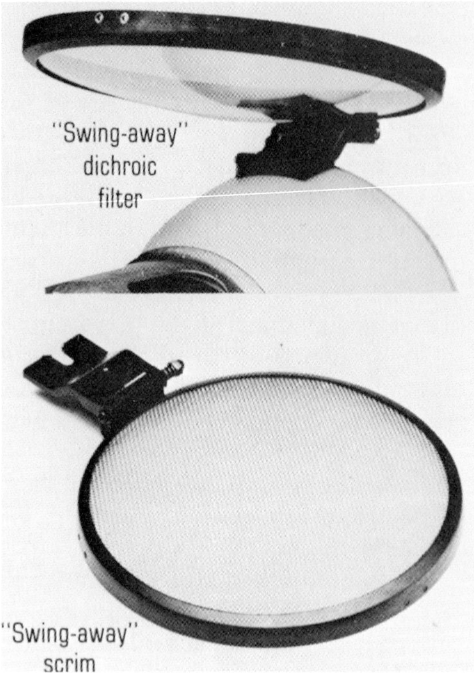

Figure 2–5 Scrim. *(Provided courtesy of Frezzolini.)*

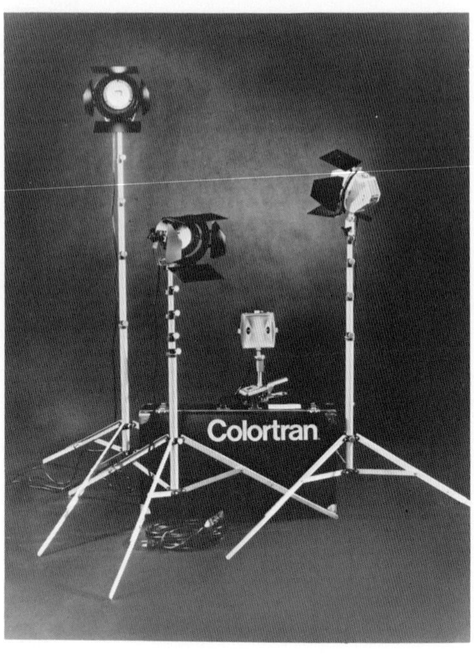

Figure 2–6A Portable light kit. *(Provided courtesy of Colortran.)*

Figure 2–6B Fixture back showing focus control. *(Provided courtesy of Colortran.)*

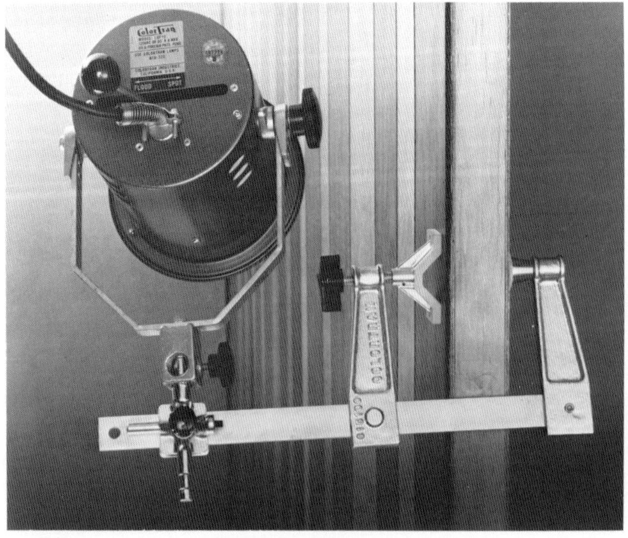

Figure 2–7A Scoop reflector. *(Provided courtesy of Colortran.)*

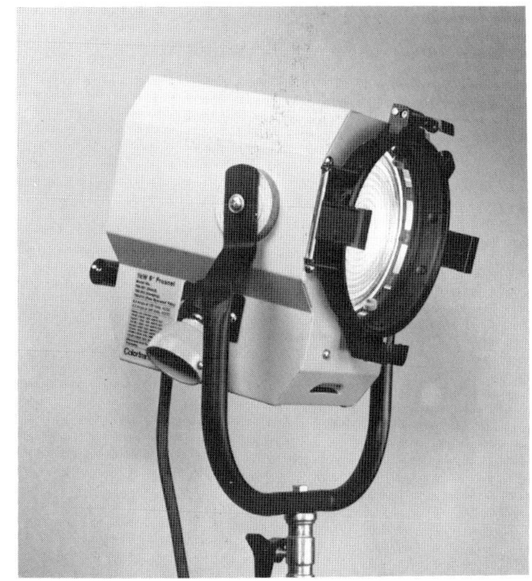

Figure 2–7B Fresnel. *(Provided courtesy of Colortran.)*

Figure 2–7C Broad floodlight. *(Provided courtesy of Colortran.)*

employ lenses in front of the lamp or vary the position of the lamp in its housing and are usually used for *key* lighting.

Reflectors are used to spread the light and provide illumination over large areas.

Concentrating lenses are used to provide *spotlighting* and to concentrate the light within smaller areas (Fig. 2–8). Spotlighting results in sharper contrasts and hard shadows on objects within the spot-

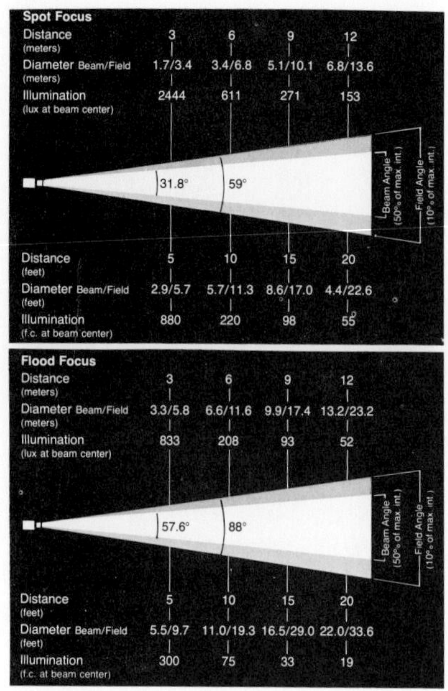

Figure 2–8 Spot focus and flood focus graphs for the portable and location lighting Mini-Pro of Colortran. *(Provided courtesy of Colortran.)*

Figure 2–9 Barn doors. *(Provided courtesy of Colortran.)*

lighted area. Often "barn doors" are clamped to the edges of the lamp fixture to adjust the field of light dispersion (Fig. 2–9).

Light-Mounting Devices

To optimize the use of a light source, it is necessary to control its placement and positioning. There are various types of lamp fixture stands, with folding tripod legs for transportable use and heavy bases for studios. There are overhead support brackets, pole extenders, pantograph suspension systems with the light fixtures counterbalanced, and mounting grids available on the high ceilings designed in the video studio for this purpose (Fig. 2–10).

Studio lighting fixtures are often wired through a control panel, so they can be switched on and off as needed from a central position rather than individually. In a studio large enough to hold two sets, the light control panel may be preprogrammed to turn on the lights of the second set just before the action switches to that set; or the panel may be used to provide differing light effects (such as day or night) on the same set (Fig. 2–11).

LIGHT PLACEMENT

Light placement has a direct effect on how the video camera sees a subject. *Flat*, or *frontal, lighting* by itself, makes the sub-

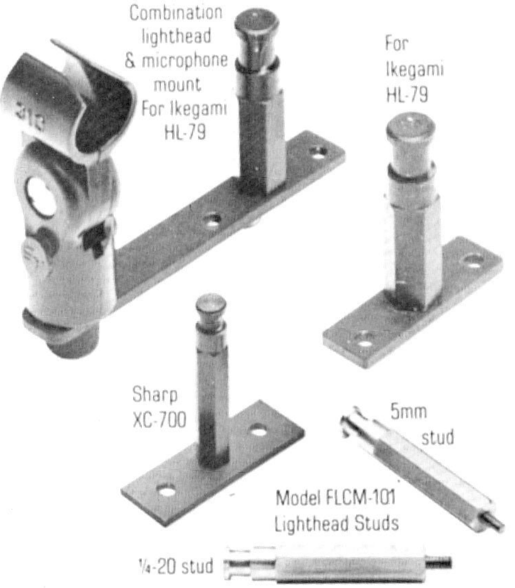

Figure 2–10A Light-mounting devices. *(Provided courtesy of Frezzolini.)*

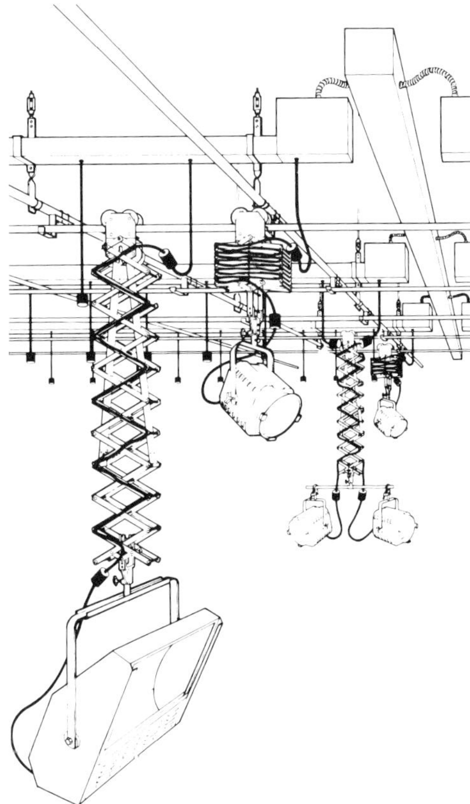

Figure 2–10B Light-mounting devices. *(Provided courtesy of Colortran.)*

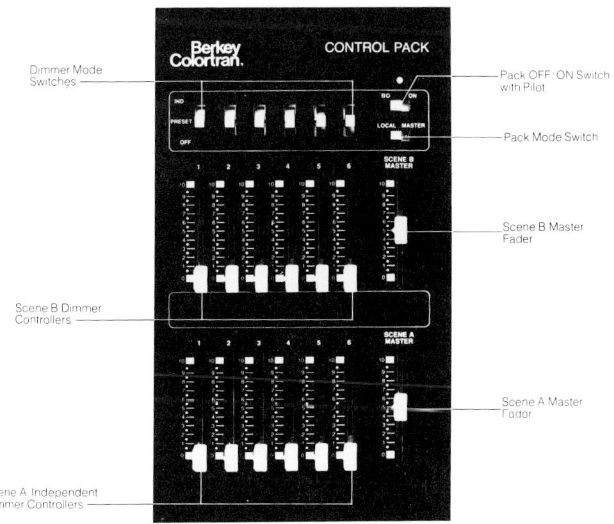

Figure 2–11 Lighting-control panel. *(Provided courtesy of Colortran.)*

ject look shadowless and without contrast, except in those places where the subject has projections or hollow spots, that is, changes in its shape (Fig. 2–12).

Side lighting illuminates one side of the subject brightly while leaving the other side of the subject unlighted. Side lighting causes projections on the subject, a nose on a face for instance, to stand out in sharp contrast, because the camera sees the projection as a highlighted area next to a very dark area. The dark area is often the unlighted side of the projection, added to the shadow of the lighted projection next to it. (See Fig. 2–13.)

Top lighting causes shadows to be cast vertically downward over the subject. Top lighting emphasizes projecting features in the same way as side lighting, reproducing them as associated areas of strong light and strong shadow (Fig. 2–14).

Figure 2–13 Side lighting photo. *(Provided courtesy of Isaac Creek.)*

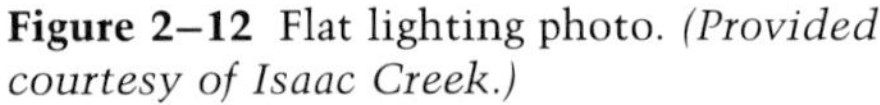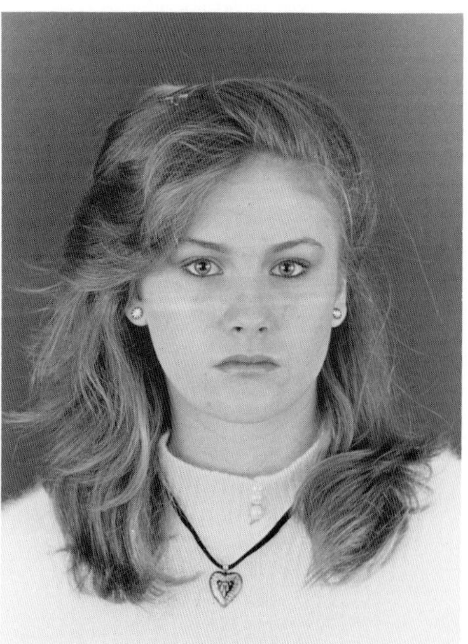

Figure 2–12 Flat lighting photo. *(Provided courtesy of Isaac Creek.)*

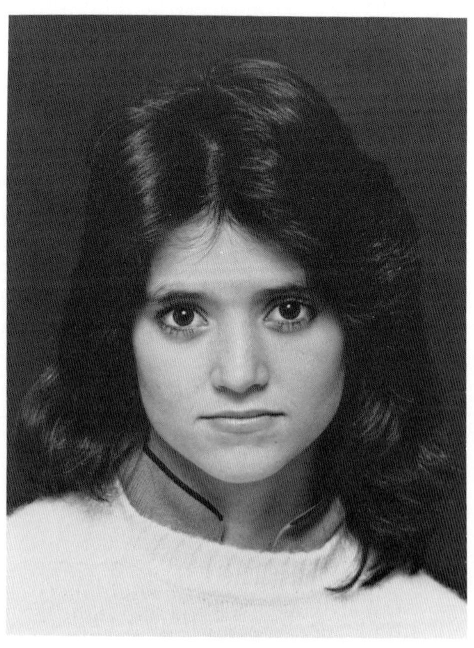

Figure 2–14 Top lighting photo. *(Provided courtesy of Isaac Creek.)*

Lighting from below causes shadows to be cast upward. Projections on the subject are lighted on their lower sides, and their upper surfaces are in shadow. The severe contrasts of lighting from below are un-natural-looking, very theatrical. Subjects appear lighted by theater stage footlights (Fig. 2–15).

Lighting from behind the subject causes all of the shadows to be cast toward the camera. The effect created is a rim of light surrounding the subject, causing small projections, like hairs, to stand out in strong contrast. Back lighting, which is often called *rim lighting*, sometimes leaves the front of the subject in complete darkness. (See Fig. 2–16.)

Lighting pointed directly at the subject is called *direct* lighting. But if the avoidance of hard shadows is desired, first the light source is aimed at a wall or low ceiling or other light-reflecting surface, and then it strikes the subject. It is called *bounce light* (Fig. 2–17).

The lighting of a scene for television, therefore, must be a combination of light placements and will include one or more of the four basic functions of a light source.

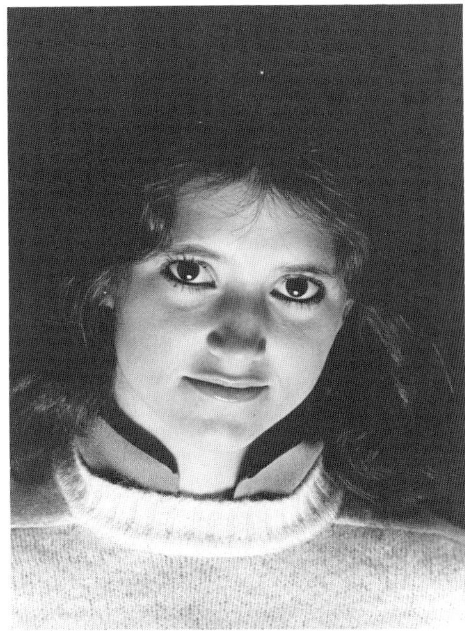

Figure 2–15 Below lighting. *(Provided courtesy of Isaac Creek.)*

A light source can be the *primary, or main,* or *modeling light,* as it is referred to in cinematography; or it can be the *key light,* as it is called in television. The light

Figure 2–16 Rim lighting photo. *(Provided courtesy of Isaac Creek.)*

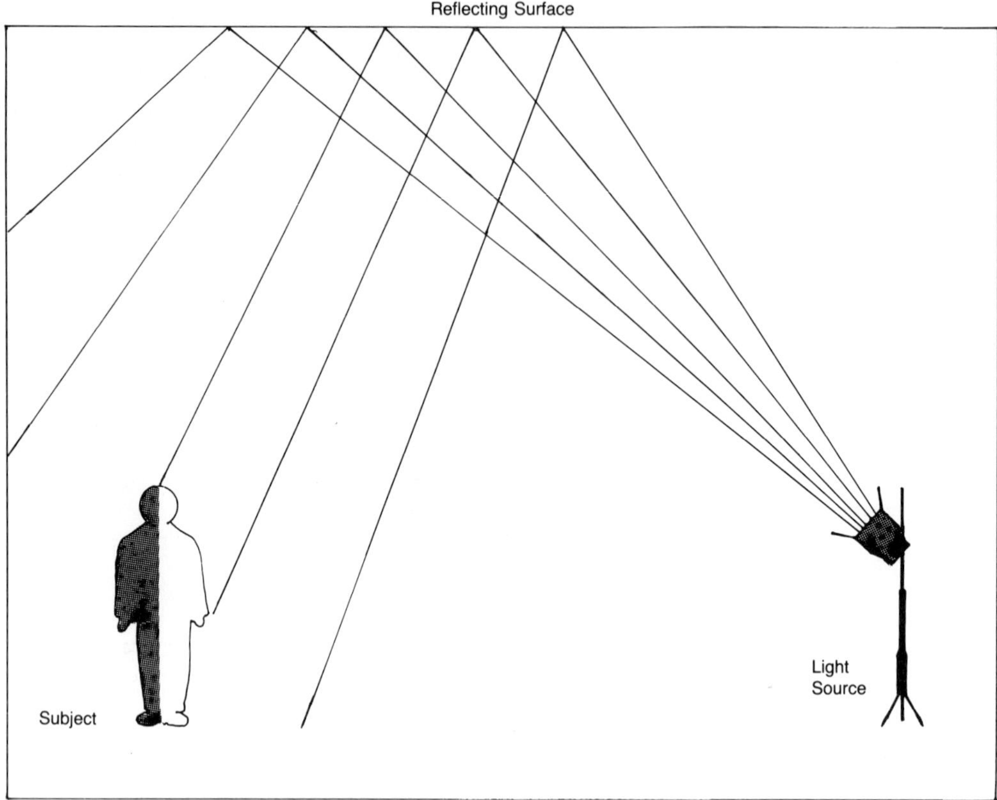

Figure 2–17 Bounce light.

source also can be the *back light,* or it can be a *fill light,* or an *effect light.*

The key light (usually a focused spotlight) is set above and normally at a 45° angle to the main subject in a scene. The back light, from behind the subject, aimed away from and at a 45° angle to the subject, separates the subject from its background. The fill light, usually angled from above and consisting of soft floodlight, provides overall illumination to the scene and ensures sufficient total light for the video camera. The effect light, if used, creates its illusion through the use of projected shadows.

In any discussion of lights, we must also discuss shadows, because shadows are in a very real sense the opposites of lights. Shadows on a subject provide contrast from light; they introduce modeling or give the subject an additional dimension. The quality of a shadow and its size and shape will vary with the area illuminated by the light source, with the size of the subject casting the shadow, and with the differences in distance between the light source, the subject, and the surface on which the shadow is cast. The quality of a shadow is described in terms of its texture and its crispness.

Often the shadows of people in a scene on the wall behind them can be distracting. To avoid this distraction, the participants are moved at least 6 feet forward from the wall. Thus the shadows fall on the floor, out of camera range, and the distraction is removed.

LIGHT MEASUREMENT

Up to this point we have treated light qualitatively. Now let us discuss it quantitatively.

Output in Lumens

The light *output* (amount of light) from an electrically operated incandescent lamp is measured in *lumens*. Here are some examples of the relationship between lamps and the number of lumens of light which they output. A 150-watt (W) light bulb produces up to 2000 lumens, a film projection lamp of 1000 W produces up to 27,000 lumens, a 10,000-W color television studio lamp produces up to 360,000 lumens, and 120 W of fluorescent lighting (three 3-foot tubes) produces about 5500 lumens.

Intensity in Footcandles and Lux

The *intensity* of light falling on a subject is measured in *footcandles* in the United States or in *lux* internationally. By convention, 1 lumen of light falling on 1 square foot of surface at a distance of 1 foot produces 1 footcandle of light intensity. And 1 footcandle of light intensity equals 10 lux, where 1 lux equals 1 lumen per meter (3 feet) squared. To put it another way, 1 lux equals the illumination per meter squared, at a distance of 1 meter, from a point source of one candle.

The important thing to remember, in the inevitable conversion from footcan-

dles to lux, is the 10-to-1 relationship: 10 lux equals 1 footcandle.

The color temperature of light is measured in degrees Kelvin. To repeat, outdoor available light is around 6400 degrees Kelvin, and electrical incandescent light varies around 3200 degrees Kelvin. The color temperature of light is measured with a color temperature meter.

Exposure Meters

The *strength*, or intensity, of light is measured with an *exposure meter*, either an incident-light meter, which reads the direct intensity of a light source, or one that provides reflectance reading of light reflected from the subject. Incident light is read with the meter *at* the subject and with the meter facing the camera. Reflected light is read with the meter *at* the camera and looking at the subject.

The most important consideration to the cameraman measuring light is how much light the camera sees. This concern may be dealt with by the use of a *spotmeter*, held directly next to the camera. Since the camera is always moved in a scene, the amount of light striking its lens system will change with each move, as measured by the spotmeter, necessitating changes in aperture settings on the camera. The spotmeter is a reflected-light meter that measures a spot on the subject to be shot by the video camera. This measurement is performed through a viewfinder having 1° of acceptance, a very narrow angle, which makes the reading very accurate. (See Fig. 2–18.)

The Minolta Spotmeter M is one example of a spotmeter. By using the combined advantages of a built-in microprocessor, memory function, and a liquid-crystal display, it determines the correct aperture setting for the camera, to create

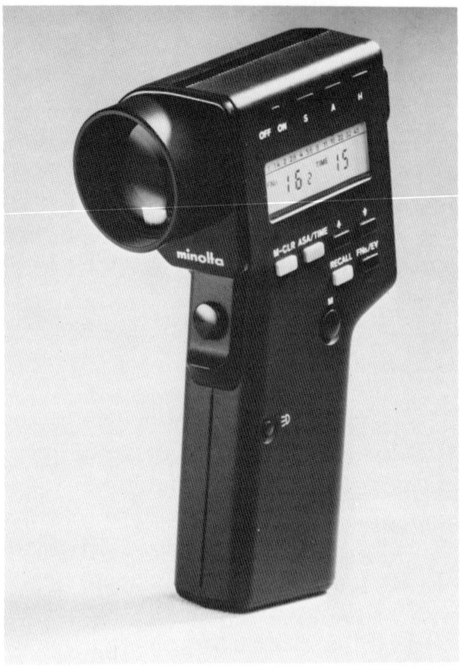

Figure 2–18 The Minolta spot meter M. *(Provided courtesy of Minolta.)*

an average, a highlight, or a shadow exposure on the spots being measured.

With each type and wattage of lamp there are associated photometric data supplied by the manufacturer. When the lamp is used as either a spotlight or a floodlight, the data include the beam angle, the field angle, the beam candlepower, and the field diameter (the number by which to multiply the distance). In this sense, lamps are rated in terms of field—wide or narrow—and throw—short, medium, or long distance. (See Fig. 2–8.)

In addition to the photometric data, the manufacturer supplies applications data such as whether the lamp fixture is best used on portable or location duty and is part of a lighting kit; whether it is best used in the studio as a spotlight, floodlight, key, back, side, or fill light.

The light that the camera "sees" is the *reflectance,* or the reflected light, from the subject. A given light source aimed at a bright-colored background will reflect more light into the camera than the same source aimed at a darker background. The texture of an object will affect its reflectance. A brown wool carpeted wall will have less reflectance than the same wall without the carpeting and painted the same color brown. The pigmentation of different paints of the same color will affect their reflectances. Regardless of the intensity of light (the amount of light) projected on a subject, the video camera sees only the light reflected from the subject.

The amount of light reflected from a subject is also related to the distance between the light source and the subject. The intensity of the light falling on a subject varies with the square of the distance between them. If the light source is moved twice as far from the subject, the intensity becomes one-fourth as strong. Conversely, halving the distance produces four times the light intensity.

THE ARTISTIC USE OF LIGHT

Light, shadow, and color projection can be used to create many dramatic effects which produce a mood, or feeling, on the part of the TV viewer. This mood creation is done with superb effect on commercially sponsored dramatic television programs which have lighting engineers to plan and set lights and the funding to provide the many lamp fixtures for the side, rim, and below lighting described earlier. The use of a sharply focused spotlight called an *ellipsoid* or a *leko* for framing areas; the projection of shadow patterns such as snowflakes, iron bars, or trees;

and the projection of Geltran colors—all these can produce many wonderful special effects.

CONTRAST RANGE

The contrast range from darkest to brightest that can be accommodated by the human eye can be assigned an arbitrary range of about 100 to 1. If we use the same frame of reference, the contrast that can be accommodated by the video camera pickup tube ranges from 20 to 1 to about 10 to 1. This relatively narrow contrast range limits the camera pickup as compared to the human eye and must always be considered by the cameraman, particularly in outdoor shooting.

Anything brighter than the camera's highest contrast limit will be compressed into peak white. And anything darker than the lowest limit, including many shadow areas easily seen by the eye, will be compressed into black and not seen by the camera.

Outdoors in bright sunlight, the contrast range can be reduced by using the neutral-density filters in the camera lens system to decrease the overall brightness (light intensity). Then the shadow areas which would have been lost can be filled in with lights or reflectors.

It is good practice to avoid shooting against very bright backgrounds, such as large expanses of bright sky and white buildings, or with the sun behind and back lighting the subject. Generally, a slight shift in angle of either the subject or the camera will enable reduction of the contrast range to a point where the camera can handle it and still take advantage of the sunlight on the front of the subject. If it becomes absolutely necessary to shoot against a sun background, such as in news gathering, then reflectors or even high-intensity light fixtures must be used to force light onto the subject to more closely match the background light and thus lower the contrast range to the camera's capacity. Failure to do so will result in a subject that has no distinguishing facial features.

In the more pragmatic terms of small-budget cable television programming, lighting is used only to illuminate the scene in the studio and to provide sufficient light to operate within the parameters of the camera pickup tube target. So it is usually limited to key, back, and fill lighting. Often two light sources will be arranged with the subject forming the apex of a triangle. This setup provides separation of the subject from the background and provides a sense of dimension.

Overhead-mounted key lighting is used in cable TV studios, and fill lighting is used to dispel strong shadows. As a rule, the back light should be twice the intensity of the key light.

For outdoor work, available light is augmented with spotlighting and fill lighting whenever necessary, mainly during evening and night work. Nighttime sports events are covered by using the lighting available at illuminated sports arenas. One attribute of a good cable TV crew member is the ability to talk the arena operator into providing additional light. Each model and make of camera has minimal light requirements for successful operation, and these must be met or exceeded for good programming.

The Lens

DEFINITION

A *lens* is an optical device made of glass and used for bending light rays by employing the natural effect of refraction. A camera lens is made by grinding a glass disk surface to its predetermined required curvature and then polishing that surface until no surface flaws are discernible.

The two surfaces of a lens element may be ground to be *concave* (curved inward), *convex* (curved outward), *flat* (with no curved surface), or a combination of these.

The curvature of the lens surface is determined by how the lens is to be used. If both lens surfaces are curved outward and the lens is fatter in the middle than at the edges, then parallel rays of light entering the front of the lens will *converge*, or bend inward, at a specific point beyond the back of the lens. If both lens surfaces are curved inward and the lens is thinner in the middle than at the edges, then parallel rays of light entering the front of the lens will *diverge*, or bend outward, from a point beyond the back of the lens. In each case, the point of convergence or divergence is called the *point of principal focus* of the lens. Lenses with different curved surfaces will provide convergence and divergence combinations based on the specific surface curvatures.

Rays of light emanating, or being reflected, from an object and entering a lens as parallel rays can be considered as a collection of individual points of light. A converging lens, by transmitting all those light points and producing point images of the object, forms an identical image of the object at a point beyond the back of the lens, the principal focus. If the target of the image pickup tube of a television camera is placed at the principal focus, then the object image will be formed at that target.

A lens may be a simple lens, composed of a single glass element, or it may be one element of a complex lens system. The individual elements of a lens system may include a combination of many different curvatures, designed to optically balance aberrations or shortcomings of single glass surfaces. Often individual lens elements are composed of two or more pieces of glass optically cemented together to form the single element. Lens systems are mounted in lens barrels which keep the lens elements separated in specific relationship to one another and have engraved markings and knurled rings on their outer surfaces for lens adjustment, discussed later.

To use a lens, we must know its *focal length*, its *speed* (or light transmittance), and its *depth of field*.

Focal Length

The focal length of a lens is described in relation to its principal focus. The principal focus of a converging lens is the point at which all of the parallel rays of light converge, beyond the back of the lens. The principal focus of a lens is a very real point, and the image that is formed at that point is focused on the video target. The distance from that image to the optical center of the lens is the focal length of the lens. It is also designed to be the "infinity" point of the lens.

The focus of a lens is controlled by turning the focus ring on the lens barrel, which in turn moves the lens elements, thus changing their interrelationship. The ring is adjusted by the cameraman until the objects are in sharp focus. The lens barrel at the focus ring has engraved distance markings, usually in both meters and feet.

Focal lengths of lenses are given in either inches or millimeters (mm), where 250 mm equals 1 inch. These are some typical focal lengths of lenses (in millimeters) for a 35-mm still camera: Under 55 mm, usually about 35 to 25 mm, lenses are called *wide-angle;* 50 to 55 mm lenses are called *normal,* or medium-focal-length; lenses between 80 and 200 mm are called *long* or *telephoto* lenses. Television camera lenses approximate those used for 16-mm movie cameras, and the normal lens for a television camera is 25 mm.

The focal length of a lens determines the size of its image capability. The longer the focal length, the larger the image capability. Thus, from a fixed point, the image formed by a 100-mm lens will be twice as large as that formed by a 50-mm lens. A so-called long lens, or one with a long focal length, will capture and enlarge a smaller field of view than a wide-angle (short focal length) lens; conversely, a wide-angle lens will capture a wider field of view.

Lens Speed

The *speed* of a lens is the factor that describes how much light will pass through the lens—its transmittance—with the lens aperture wide open. Some lenses are "faster" than others. The aperture opening is expressed as an *f*-stop number, which is the ratio of the size of the lens opening to the focal length. The range of *f*-stops of a lens is engraved on the lens barrel. The smallest *f*-number on the lens with the aperture wide open is the lens speed. Some typical lens speeds marked on a zoom-lens barrel are *f*/1.8, *f*/2.8, *f*/4, *f*/5.6, *f*/8, *f*/11, *f*/16, and *f*/22. This array of *f*-stops indicates that the lens aperture may be opened up from *f*/22 to *f*/1.8. Conversely stated, the lens aperture may be stopped down from *f*/1.8 to *f*/22. The lens speed is adjusted by the cameraman by turning a ring on the lens barrel, which internally adjusts the size of the hole, or aperture, in the center of the iris diaphragm. The *aperture* is a circular, variable-sized door behind the lens. The iris opens as wide as the lens itself, but never

closes completely on a still or movie camera. On a video camera, the aperture does close all the way, to protect the pickup tube. The *f*-number refers to the size of the hole. The larger the *f*-number, the smaller the hole. As the *f*-numbers progress in one direction, each number marking, or *stop*, doubles the size of the hole, or aperture. In the other direction, each *f*-stop halves the size of the aperture. When the size of the aperture is doubled (moved to the next highest number), twice as much light can come through the lens. When the aperture is halved (moved to the next lowest number), half as much light can come through the lens. Making the aperture smaller is called *stopping down*; conversely, making it larger is called *opening up*.

Depth of Field

By definition, a lens is supposed to be able to form a sharp image at one specific distance, with objects at other distances, either nearer or farther, being out of focus and not sharp. Finding that distance is called *focusing the lens*.

In truth, however, there is a zone of acceptable sharpness, since the human eye is not a perfect measuring device. The length of this zone, from front to back, is called the *depth of field* of a lens. The field behind the optimal focus point is usually deeper than the field in front.

The depth of field can be increased by stopping down (making the aperture smaller) or by focusing the lens at an object that is farther away. Closing the aperture, however, generally will require adding light to the subject in indoor shooting. The depth of field is determined by the aperture opening and the lens-to-subject distance. The depth of field can be decreased by opening up (making the aperture larger) or by focusing the lens on an object that is nearer.

Often lens surfaces are coated, to prevent lens surface reflection; are shaded, with an attachment called a *lens shade*, to prevent glare from very bright sources near the object of focus or to prevent raindrops from hitting the lens; and are capped when not in use, to avoid lens surface dirt and to keep unwanted light from entering the video camera when no video picture is required.

ZOOM LENSES

In the earlier days of television, as many as four separate lenses of differing fixed focal lengths were mounted on a revolving turret on the front of the video camera. The cameraman chose on his own, or was instructed by the director, to go (turn) to a particular lens for a specific shot. The cameraman chose the lens by turning a handle at the back of the camera, which mechanically turned the turret at the front of the camera. The problem was that the lens always seemed to be in the process of being changed when the camera was switched or faded on the air. Some dreadful television production effects resulted.

Individual lenses in a turret have been replaced in television by a single multipurpose lens which has the property of *zooming*, or continuously changing the image size or field of view of an object without moving the camera. The zoom lens has a variable focal length. It is a multielement lens whose elements are made to move within the barrel in relationship to one another by the use of complex cams and followers. Theoretically, the zoom lens is not as sharp as individual lenses; but, theory aside, the zoom lens is accepted by the television industry as its standard.

Zoom lenses are available in focal lengths ranging from 7 to over 1000 mm and with apertures as wide as *f*/1.6. Manufacturers' ratings of their lenses are marked on the lens barrel as zoom range × focal length. Thus a 15×9 lens has a zoom ratio of 15:1 and a *minimum* focal length of 9 mm. The focal length is always given at the minimum rating. Typical zoom ratios found are 3.5:1, 10:1, 12:1, 14:1, 15:1, 16:1, 17:1, 25:1, 28:1, and 30:1. This capability for wide variation in focal length enables the cameraman to achieve great scene variation without camera movement. The minimum-to-maximum focal length of a zoom lens is called its *zoom range*. (See Fig. 2–19.)

All modern television cameras employ zoom lenses. The studio cameraman may use any of the three lens controls at his end of the camera that are mounted on the pan-tilt handle. These controls are the zoom control, the aperture control, and the focus control. On some cameras, the aperture control is automatic, and in studio use aperture control is often left to the technician who remotely controls all of the cameras in the control room. The technician views the video picture on a larger monitor than the cameraman and has a vectorscope and waveform monitor

Figure 2–19 Zoom lens diagram. *(Provided courtesy of Angenieux.)*

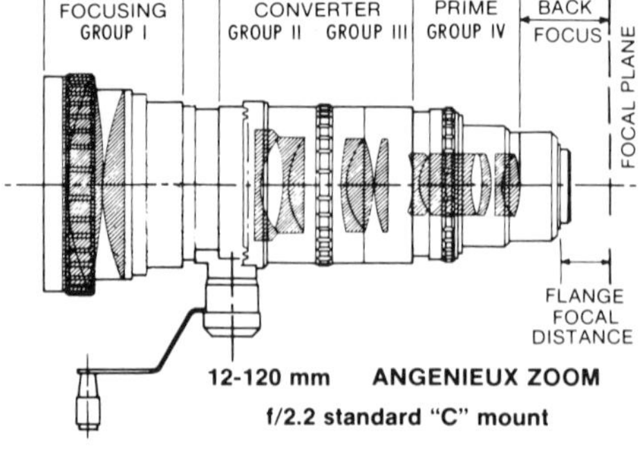

at his disposal as well. This technician is known as the *video control man.*

Television camera pickup tubes have image formats (target sizes) of ½ inch, ⅔ inch, 1 inch, 1.25 inches, and 1.5 inches. The camera lens size varies with the pickup tube target size, which in turn is determined by the type of pickup tube used. The zoom ratio is determined by whether the camera is primarily an indoor or outdoor camera, since outdoor cameras, unrestricted by studio walls, can use much larger lenses.

The zoom lens format (size at the camera end) must complement the target size, and so video zooms are made in ⅔ inch, 1 inch, and 1.25 inches, which handle all the pickup tube sizes.

Of primary importance to the video cameraman is the field of view that will be available with a given zoom lens. Figure 2–20 shows field-of-view tables for the three zoom lens formats.

Focus

Focusing a zoom lens is a bit more complex than focusing a lens with a fixed focal length. It should be done in three steps. First, zoom (move the zoom control, or turn the zoom ring) to the longest focal length, or telephoto setting, at full aperture. That is, zoom to a tight shot. *Zoom in.* Second, focus sharply, with the focus control or ring. Third, zoom back, or *Zoom out,* to a wider angle or to where you want to frame the object in the picture.

An object is in most critical focus at the longest (telephoto or tight) lens setting of the zoom lens. If the object is in focus at that point, it should stay in focus all the way in the opposite direction to a wide-angle shot.

When the object is in motion and changes the focal plane, two things must be done simultaneously to keep the object in focus. The cameraman uses one hand to zoom and the other hand to adjust for sharp focus. This is often called *pull focus,* a term retained from motion picture filming. Once the focus has been changed on a moving object, the zoom lens is no longer in focus for its full range. At the first opportunity after the subject has stopped moving, the zoom lens must be refocused by zooming in tight on the next subject and refocusing at that focal plane.

The smoothest zooms are performed by using a *servomotor* to turn the zoom ring on the lens barrel. Servomotor zooms have a damping effect, allowing the operator to start the zoom slowly and then accelerate to the desired speed, with a reversal at the end of the zoom, making the zoom itself less noticeable to the viewer. The servomotor control is a rocker switch that is found adjacent to the lens and operated with two fingers. It is marked W (wide) on one side, and T (tight or telephoto) on the other. The control is rocked forward to zoom to a tight shot and is pressed hard for a fast zoom or gently for a slow zoom. The control is rocked back for a wide-angle shot, again, gently for a slow zoom and hard for a fast zoom.

Back Focus

An adjustment called *back focus* must be made whenever the plane of the back of the zoom lens is changed in relation to the face of the pickup tube. This happens whenever a pickup tube is replaced, the lens is removed and replaced on the camera, or the camera is jolted sufficiently to change the plane relationship. The adjustment is made on the inside of the camera by moving the position of the pickup tube.

To back-focus the lens, zoom *in* to the

FIELD OF VIEW — 2/3" TUBE (8.8 x 6.6 mm)

HORIZONTAL & VERTICAL — FEET & INCHES (Rounded off)
CAMERA TO SUBJECT DISTANCE IN FEET

EFL in mm	2'	3'	5'	10'	15'	20'	30'
	HxV	HxV	HxV	HxV	HxV	HxV	HxV
4	4'5" x 3'4"	6'7" x 4'11"	11' x 8-1/4'	22' x 16-1/2'	33' x 24-3/4'	44' x 33'	66' x 49-1/2'
7	2'6" x 1'10"	3-3/4' x 2'10"	6-1/4' x 4'8"	12-1/2' x 9-1/2'	19' x 14'	25' x 19'	38' x 28'
9	2' x 1-1/2'	2'11" x 2'2"	4'11" x 3'8"	9'9" x 7'4"	14'8" x 11'	19'7" x 14'8"	29'4" x 22'
9.5	1'10" x 1'5"	2-3/4' x 2'	4'7" x 3'5"	9' x 7'	14' x 10-1/2'	18-1/2' x 14'	28' x 21'
12	1-1/2' x 1'	2'2" x 1'8"	3'8" x 2'9"	7'4" x 5'6"	11' x 8-1/4'	14-3/4' x 11'	22' x 16-1/2'
15	1'2" x 10-1/2"	1'9" x 1'4"	2'11" x 2'2"	5'10" x 4'4"	8'10" x 6'7"	11'9" x 8'10"	17-1/2' x 13-1/4'
20	10" x 8"	1'4" x 12"	2'2" x 1'8"	4'5" x 3'4"	6-1/2'x 5'	8'10" x 6'7"	13-1/4' x 9-3/4'
25	8-1/2" x 6-1/2"	1'1" x 9-1/2"	1'9" x 1'4"	3'6" x 2'7"	5'3" x 3'2"	7' x 5'4"	10'7" x 8'
40	5-1/4" x 4"	8" x 6"	1'1" x 10"	2'2" x 1'8"	3'4" x 2'6"	4'5" x 3'4"	6'7" x 4'11"
50	4-1/4" x 3-1/8"	6-3/8" x 4-3/4"	10-1/2" x 8"	1'9" x 1'4"	2'8" x 2'	3-1/2' x 2'7"	5'3" x 4'
60	3-1/2" x 2-5/8"	5-1/4" x 4"	8-7/8" x 6-5/8"	1'6" x 1'1"	2'2" x 1'8"	3' x 2'2"	4'5" x 3'4"
75	2-7/8" x 2-1/8"	4-1/4" x 3-1/8"	7" x 5-1/4"	1'2" x 10-1/2"	1'9" x 1'4"	2'4" x 1'9"	3'6" x 2'7-1/2"
100	2-1/8" x 1-5/8"	3-1/8" x 2-3/8"	5-1/4" x 4"	10-1/2" x 8"	1'4" x 1'	1'9" x 1'4"	2'8" x 2'
120	1-3/4" x 1-3/8"	2-5/8" x 2"	4-3/8" x 3-1/4"	8-3/4" x 6-5/8"	1'1-1/4" x 9-7/8"	1'5-5/8" x 1'1-1/4"	2'2-3/8" x 1'7-3/4"
142	1-1/2" x 1-1/8"	2-1/4" x 1-5/8"	3-3/4" x 2-3/4"	7-1/2" x 5-5/8"	11'1/8" x 8-3/8"	1'3" x 11-1/4"	1'10-1/4" x 1'4-3/4"
180	1-1/8" x 7/8"	1-3/4" x 1-3/8"	3" x 2-1/8"	6" x 4-3/8"	8-3/4" x 6-5/8"	11-3/4" x 8-3/4"	1'6" x 1'1-1/4"
225	1" x 3/4"	1-3/8" x 1"	2-3/8" x 1-3/4"	4-5/8" x 3-1/2"	7" x 5-1/4"	9-3/8" x 7"	1'2" x 10-1/2"
240	7/8" x 5/8"	1-1/4" x 7/8"	2-1/4" x 1-5/8"	4-3/8" x 3-1/4"	6-5/8" x 5"	8-3/4" x 6-5/8"	1'1/4" x 9-7/8"
400	1/2" x 3/8"	3/4" x 5/8"	1-3/8" x 1"	2-5/8" x 2"	4" x 3"	5-1/4" x 4"	8" x 6"
615	3/8" x 1/4"	1/2" x 3/8"	7/8" x 5/8"	1-3/4" x 1-1/4"	2-5/8" x 2"	3-1/2" x 2-5/8"	5-1/8" x 3-7/8"

EFL-Effective Focal Length

Figure 2–20A Fields of view for ⅔-inch tube. *(Provided courtesy of Angeniux.)*

FIELD OF VIEW — 1" TUBE (12.8 x 9.6mm)
HORIZONTAL & VERTICAL - FEET & INCHES (Rounded off)
CAMERA TO SUBJECT DISTANCE IN FEET

EFL in mm	3'	5'	10'	15'	20'	30'	50'	100'
	HxV	HxV	HxV	HxV	HxV	HxV	HxV	HxV
9.5	4' x 3'	6'9" x 5'3/4"	13'6" x 10'1"	20'2" x 15'2"	26'11" x 20'2"	40'5" x 30'4"	67'5" x 50'6"	137'9" x 100'
12.5	3'1" x 2'4"	5'1" x 3'10"	10'3" x 7'8"	15'4" x 11'6"	20'6" x 15'4"	30'9" x 23'	51'2" x 38'5"	102'5" x 76'10"
14	2'9" x 2'1"	4'7" x 3'5"	9'2" x 6'10"	13'8" x 10'3'	18'3" x 13'9"	27'5" x 20'7"	45'8" x 34'4"	91'5" x 68'7"
20	1'11" x 1'5"	3'2" x 2'5"	6'5" x 4'10"	9'7" x 7'2"	12'10" x 9'7"	19'2" x 14'5"	32' x 24'	64' x 48'
24	1'7" x 1'3"	2'8" x 2'	5'4" x 4'	8' x 6'	10'8" x 8'	16' x 12'	26'8" x 20'	53'4" x 40'
30	1'3" x 11-1/2"	2'2" x 1'7"	4'3" x 3'2"	6'5" x 4'10"	8'6" x 6'5"	12'10" x 9'7"	21'4" x 16'	42'8" x 32'
50	9-1/4" x 6-7/8"	1'3" x 11-1/2"	2'7" x 1'11"	3'10" x 2'11"	5'1" x 3'10"	7'8" x 5'9"	12'10" x 9'7"	25'7" x 19'2"
79	5-7/8" x 4-3/8"	9-3/4" x 7-3/8"	1'7" x 1'3"	2'5" x 1'10'	3'3" x 2'5"	4'10" x 3'8"	8'1" x 6'1"	16'2" x 12'2"
100	4-5/8" x 3-1/2"	7-5/8" x 5-3/4"	1'3" x 11-1/2"	1'11" x 1'5'	2'7" x 1'11"	3'10" x 2'11"	6'5" x 4'10"	12'10" x 9'7"
140	3-1/4" x 2-1/2"	5-1/2" x 4-1/8"	11" x 8-1/4"	1'4" x 1'	1'10" x 1'4"	2'9" x 2'	4'7" x 3'5"	9'2" x 6'10"
150	3" x 2-3/8"	5-1/8" x 3-7/8"	10-1/4" x 7-5/8"	1'3" x 11-1/2"	1'9" x 1'3"	2'7" x 1'11"	4'3" x 3'2"	8'6" x 6'5"
188	2-1/2" x 1-7/8"	4" x 3"	8-1/8" x 6-1/8"	1'0" x 9"	1'4" x 1'	2' x 1'6"	3'5" x 2'7"	6'10" x 5'1"
200	2-1/4" x 1-3/4"	3-7/8" x 2-7/8"	7-5/8" x 5-3/4"	11-1/2" x 8-5/8"	1'3" x 11-1/2"	1'11" x 1'5"	3'2" x 2'5"	6'5" x 4'10"
210	2-1/8" x 1-5/8"	3-5/8" x 2-3/4"	7-3/8" x 5-1/2"	11" x 8-1/4"	1'2-5/8" x 11"	1'10" x 1'4"	3'0" x 2'3"	6'1" x 4'7"
300	1-1/2" x 1-1/8"	2-1/2" x 2"	5-1/8" x 3-7/8"	7-5/8" x 5-3/4"	10-1/4" x 7-5/8"	1'3" x 11-1/2"	2'2" x 1'7"	4'3" x 3'2"
400	1-1/8" x 7/8"	2" x 1-1/2"	3-7/8" x 2-7/8"	5-3/4" x 4-3/8"	7-5/8" x 5-3/4"	11-1/2" x 8-5/8"	1'7" x 1'2"	3'2" x 2'5"
525	7/8" x 5/8"	1-1/2" x 1"	3" x 2-1/4"	4-3/8" x 3-1/4"	5-7/8" x 4-3/8"	8-3/4" x 6-5/8"	1'3" x 11"	2'5" x 1'10"
800	5/8" x 3/8"	1" x 3/4"	2" x 1-1/2"	2-7/8" x 2-1/8"	3-7/8" x 2-7/8"	5-3/4" x 4-3/8"	9-5/8" x 7-1/4"	1'7" x 1'2"
1000	1/2" x 3/8"	3/4" x 1/2"	1-1/2" x 1-1/8"	2-3/8" x 1-3/4"	3" x 2-1/4"	4-5/8" x 3-1/2"	7-5/8" x 5-3/4"	1'3" x 11-1/2"

EFL-Effective Focal Length

Figure 2–20B Field of view for 1-inch tube. *(Provided courtesy of Angenieux.)*

FIELD OF VIEW — 1¼" TUBE (17.1 x 12.8mm)

HORIZONTAL & VERTICAL - FEET & INCHES (Rounded off)
CAMERA TO SUBJECT DISTANCE IN FEET

EFL	3'	5'	10'	15'	20'	30'	50'	100'
16	3'2" x 2'4"	5'4" x 4'	10'8" x 8'	16' x 12'	21'4" x 16'	32' x 24'	53'5" x 40'	106'10" x 80'
18	2'10" x 2'1"	4'9" x 3'6"	9'6" x 7'1"	14'3" x 10'8"	19' x 14'2"	28'6" x 21'4"	47'6" x 35'6"	95' x 71'1"
25	2' x 1'6"	3'5" x 2'6"	6'10" x 5'1"	10'3" x 7'8"	13'8" x 10'2"	20'6" x 15'4"	34'2" x 25'7"	68'4" x 51'2"
27.5	1'10" x 1'4"	3'1" x 2'4"	6'2" x 4'7"	9'3" x 6'11"	12'4" x 9'3"	18'7" x 13'11"	31' x 23'3"	62'2" x 46'6"
50	1' x 9"	1'8" x 1'3"	3'5" x 2'6"	5'1" x 3'10"	6'10" x 5'1"	10'3" x 7'8"	17'1" x 12'9"	34'2" x 25'7"
70	8-3/4" x 6-1/2"	1'2" x 10-7/8"	2'5" x 1'10"	3'8" x 2'9"	4'10" x 3'7"	7'4" x 5'5"	12'2" x 9'1"	24'5" x 18'3"
100	6-1/8" x 4-1/2"	10-1/4" x 7-5/8"	1'8" x 1'3"	2'6" x 1'11"	3'5" x 2'6"	5'1" x 3'10"	8'6" x 6'4"	17'1" x 12'9"
145	4-1/8" x 3-1/8"	7" x 5-1/2"	1'2" x 10-1/2"	1'9" x 1'3"	2'4" x 1'9"	3'6" x 2'7"	5'10" x 4'5"	11'9" x 8'10"
160	3-3/4" x 2-7/8"	6-3/8" x 4-3/4"	1' x 9-1/2"	1'7" x 1-1/2"	2'1" x 1'7"	3'2" x 2'4"	5'4" x 4'	10'8" x 8'
180	3-3/8" x 2-1/2"	5-5/8" x 4-1/4"	11-3/8" x 8-1/2"	1'5" x 1'	1'10" x 1'5"	2'10" x 2'1"	4'9" x 3'6"	9'6" x 7'1"
200	3" x 2-1/4"	5-1/8" x 3-3/4"	10-1/4" x 7-5/8"	1'3" x 11-1/2"	1'8" x 1'3"	2'6" x 1'11"	4'3" x 3'2"	8'6" x 6'4"
300	2" x 1-1/2"	3-3/8" x 2-1/2"	6-3/4" x 5-1/8"	10-1/4" x 7-5/8"	1'1" x 10-1/4"	1'8" x 1'3"	2'10" x 2'1"	5'8" x 4'3"
400	1-1/2" x 1-1/8"	2-1/2" x 1-7/8"	5-1/8" x 3-3/8"	7-5/8" x 5-3/4"	10-1/4" x 7-5/8"	1'3" x 11-1/2"	2'1" x 1'7"	4'3" x 3'2"
675	7/8" x 5/8"	1-1/2" x 1-1/8"	3" x 2-1/4"	4-1/2" x 3-3/8"	6" x 4-1/2"	9" x 6-3/4"	1'3" x 11-3/8"	2'6" x 1'10"
750	3/4" x 1/2"	1-1/4" x 1"	2-5/8" x 2"	4" x 3"	5-3/8" x 4"	8-1/8" x 6-1/8"	1'1" x 10-1/8"	2'3" x 1'8"
1000	5/8" x 3/8"	1" x 3/4"	2" x 1-1/2"	3" x 2-1/4"	4" x 3"	6-1/8" x 4-1/2"	10-1/4" x 7-5/8"	1'8" x 1'3"
3000	3/16" x 1/8"	5/16" x 1/4"	5/8" x 1/2"	1" x 3/4"	1-1/4" x 1"	2" x 1-1/2"	3-3/8" x 2-1/2"	6-3/4" x 5-1/8"

EFL-Effective Focal Length

Figure 2–20C Field of view for 1¼-inch tube. (*Provided courtesy of Angenieux.*)

longest focal length and focus sharply on an object. Then zoom *out* to the shortest focal length (the wide angle) and adjust the planar position of the pickup tube. Next, refocus and zoom to telephoto again. Now the lens will be in back focus in relation to the particular plane of the tube face, throughout its zoom range.

The same rules regarding depth of field apply to both zoom lenses and fixed lenses. If you zoom all the way out to the shortest focal length, there will be maximum depth of field; if you zoom all the way in to a tight shot, the minimum depth of field will result.

When any movement of the subject is expected, the cameraman should avoid being zoomed in, to telephoto, where the subject can easily move out of the picture by moving a few inches. Also, *camera* movement of any kind during zoomed in to telephoto will cause exaggerated picture movement.

Long-focal-length settings seemingly push objects together front to back or spread them apart on short-focal-length settings.

A peculiarity of the zoom lens is image centering. Because of the changing magnification during zooming, any point in the image that is not on axis will move proportionally away from axis as magnification increases, or as the focal length increases. To maintain good centering throughout the zoom range, start by centering the object at the longest focal length of the lens. See Fig. 2–21.

Zoom lenses should be treated as fragile objects. The front lens element should be cleaned before each use with blown air only, never with a lens brush or a handkerchief. This is another of those very valid admonitions offered by lens manufacturers that are observed primarily by misuse.

REVIEW QUESTIONS

1. Define light.
2. What happens to light rays when they hit an object?
3. Describe the types of light instruments used in television.

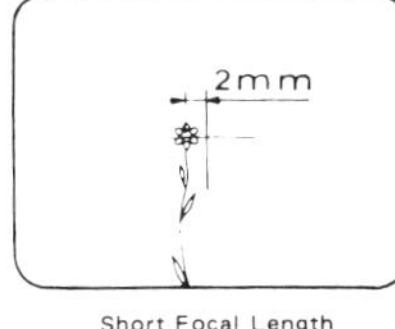

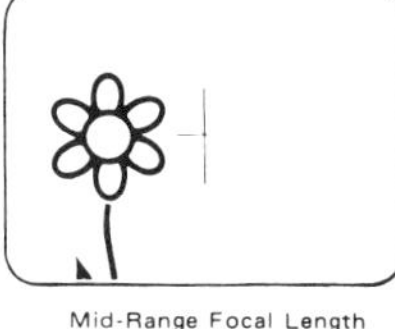

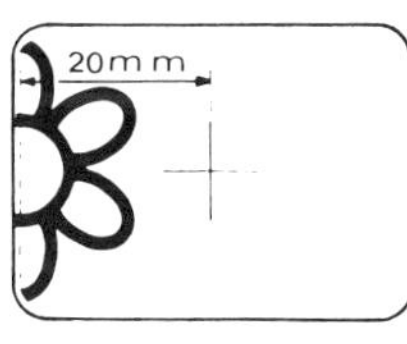

Figure 2–21 Object centered and object not centered diagrams. *(Provided courtesy of Angenieux.)*

In order to maintain good centering throughout the zoom range of a lens, it is advisable to initially center a subject at the longest focal length.

CENTERED (on optical axis)

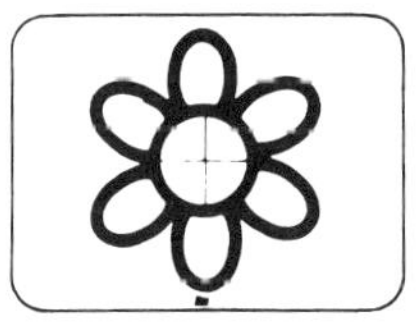

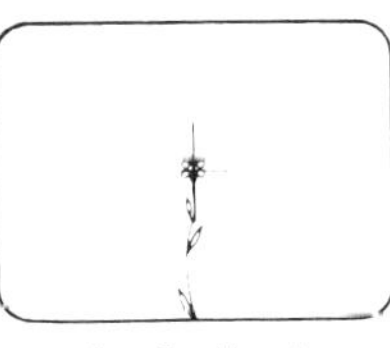

Figure 2–22 Typical zoom lens. *(Provided courtesy of Angenieux.)*

4. What is the most prevalent lamp type?

5. Describe a scrim. A barn door.

6. How are light fixtures mounted?

7. Describe side, bounce, key, rim, and below lighting. What effects does each cause?

8. How is light measured? What is light output measured in? What measurement is used for intensity?

9. What is a lens?

10. Discuss the speed, depth of field, and focal length of a lens.

11. Describe a zoom lens. How is a zoom lens focused?

12. What is back focus?

Chapter 3

The Video Camera

THE CAMERA

A camera, any camera, is a lighttight oblong box with a lens at one end of the oblong and a light-sensitive surface at the other end. The distance between the back of the lens and the light-sensitive surface is the principal focus of the lens and the back focus of a zoom lens.

In a film camera, after light has entered the lens and struck the sensitive film, the door that allowed the light to enter, called the lens *shutter,* is closed, and the film is advanced to the next frame. The film is advanced slowly, and the shutter is opened and closed intermittently by the camera operator using a *still* camera; the film is advanced rapidly and the shutter opened and closed quite rapidly for a movie camera.

In a video camera, light enters the lens and strikes the sensitive area of the pickup tube, called the *target area,* which is located behind glass at the front of the tube (See Fig. 3–2). The target area is made of a photoconductor metal alloy. For instance, the Saticon tube target is made of selenium (S), *arsenic* (A), and *tellurium* (T). Vidicons use antimony trisulfide, and Plumbicons and Leddicons use oxides of lead. Pickup tube targets are extremely sensitive to light. A brief moment of in-

Figure 3–1 Video camera. *(Provided courtesy of JVC.)*

tense light can ruin a target. Targets also have a finite life span that is related directly to the amount of light they have "seen." It is customary, therefore, to keep lenses capped or to protect targets from light whenever they are not in use, to extend life. (See Fig. 3–2.) The target area is stationary; it does not move. There is no shutter in a video camera, but there is an iris which may be closed.

Instead of movement of the target (like film advance movement), there is a cathode-ray scanning circuit in the camera. This circuit operates the pickup tube's deflection, or sweep, system, which is co-

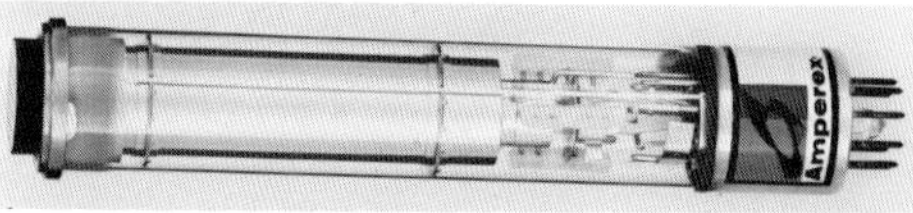

Figure 3–2 Video camera pickup tube. *(Provided courtesy of Amperex.)*

ordinated with external electronic sweep circuits. These circuits move an electron beam back and forth across the front of the pickup tube, pick up the picture information, and clear, or wipe, the surface of the pickup tube target a given number of times per second, preparing that surface each time for a new exposure. Each exposure is called a *frame*, and each frame is composed of two *fields*. One field is composed of all the odd-numbered sweep lines in the frame, and the other field is composed of all the even-numbered sweep lines. (See Fig. 3–3.)

The number of lines in the two fields and the sweep frequency are determined by the type of system being discussed. Several systems are in use, with the most prevalent in the United States being the National Television Standards Committee (NTSC) system, which employes 525 lines and a sweep frequency of 60 fields (30 frames) per second. The NTSC system is required in U.S. broadcast television by the Federal Communications Commission, television's regulatory body. Other systems include the PAL and SECAM systems used in many other countries.

The relationship of the picture height to the picture width is standardized in an aspect ratio of 3:4 in U.S. television.

Clearly the video camera system is a bit more complex than a film system. Let us look at how the video signal is put together by the video camera and its associated equipment. Light, upon entering

Figure 3–3 Frame and fields diagram.

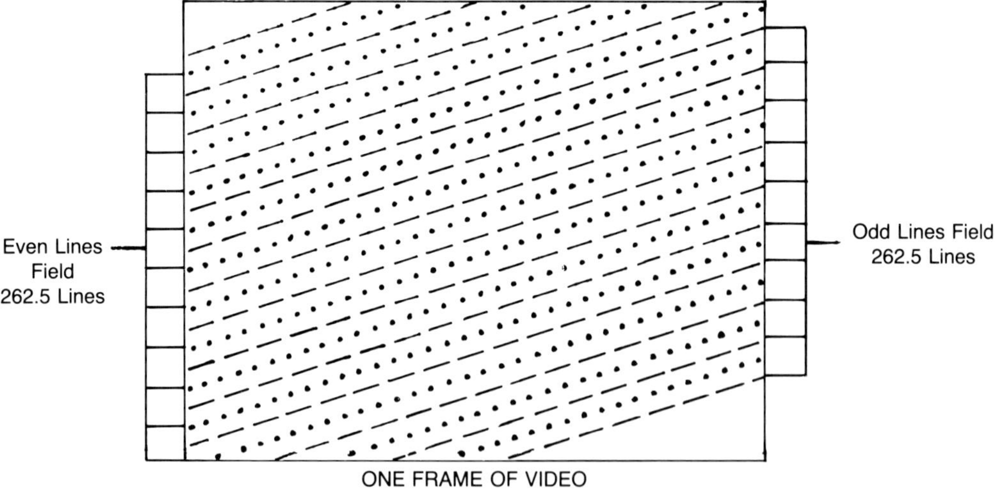

a video camera, is transformed to electronic information in specific patterns; and identical electronic information patterns, when they control the picture tube of a television set, are transformed back again to light. Both the camera pickup tube and the TV picture tube are known as *transducers*, which are devices that can change one form of energy to another form of energy. The video picture, then, travels throughout the video system as pulses and patterns of electricity.

Another factor to consider at this point is how the human eye treats pictures. If we watch *still* pictures that change rapidly, the eye perceives this as motion. And if the pictures change at a rate of 24 to 30 times each second or, as we described it earlier, at a rate of 30 *frames* each second, then the eye sees this as continuous motion. The figure of 30 frames per second was chosen as the U.S. standard because 30 is a submultiple of the 60-hertz (Hz) alternating-current (ac) power line frequency that is standard throughout the United States.

FRAMES AND FIELDS

The NTSC video system uses a frame change of 30 times per second, with each frame having 525 horizontal scanning lines, starting at the top of the picture and descending to the bottom. Since the top lines would fade before the bottom lines could be scanned, causing the picture to flicker, the frames are divided into two fields, each with 262.5 lines, and the lines in the two fields are interlaced. The first field scans the odd-numbered lines, and the second field scans the even-numbered interlaced lines.

The scanning is done by a beam of electrons created within the transducer. In its natural state, the beam would appear as a narrow stream of water emitted by a hose, hitting the center of wherever it is aimed.

To make the electron beam move from side to side, and again from bottom to top of the picture, magnetic electronic beam deflectors are used. These deflectors can be either coils or plates, which are mounted at the neck of the transducer and which magnetically move the beam in a predetermined, carefully timed program.

The timing is controlled by pulses of voltage, and the pattern of light that is traced by the pulse-controlled electron beam of a receiver or video monitor picture tube is called a *raster*. The raster is composed of the horizontal scanning lines, which in turn are the sweep of the electron beam from left to right across the face of the transducer.

SYNC AND BLANKING

Each horizontal line is timed to last 63.5 microseconds, and there are 15,734 of them in each second. When the beam reaches the right side of the picture, it is *blanked out* by a horizontal blanking pulse for 10.45 microseconds so that it can retrace its path from right to left, to begin the next line. During this time period, when the video is blanked out, a horizontal synchronizing (sync) pulse lasting 4.7 microseconds is added to the signal. This pulse controls the timing of the horizontal deflection circuits, which sweep the electron beam from side to side. Following the horizontal sync pulse there is added a short color burst of 8 cycles of 3.58 megahertz (MHz) color subcarrier which synchronizes the color information for the upcoming horizontal line of the picture.

When the beam has reached the bottom

of the raster and has traced the last horizontal line in the field, it is again *blanked out*, or turned off, this time for 1333 microseconds. This is called the *field blanking interval*, the *vertical blanking interval*, or simply the *vertical interval*. The vertical interval takes the equivalent time of 21 horizontal lines, during which time the beam returns to the top of the raster. During this vertical-interval period, some very important timing, or scheduling, information is added to the video signal. To the time that would be used by the first 9 of the 21 lines, equalizing pulses and the vertical sync pulse are added, which control the timing of the vertical beam deflection circuits. The time for the next 6 lines (lines 10 through 16) is unused and held for possible future use. The vertical-interval test signals (VITS), which are employed for quality control by the television networks, occupy the time of lines 17 and 18. Line 19 time contains the vertical-interval reference signal (VIRS), which is used to control the picture color and level. And the time for lines 20 and 21 is used to carry nonprogram information such as video text services. Line 22 is the first line of the next picture.

All of the synchronizing pulses are generated or controlled by a device usually external to the camera, but closely associated with it, called a *sync generator*. Then the *composite video*, made up of the picture and sync information combined, can be "channelized"; that is, placed on a carrier frequency channel which is 6 MHz wide. Of that 6 MHz, the composite video takes up 4.5 MHz, and *within* that 4.5 MHz there is the 3.58-MHz color subcarrier. The rest of the 1.5-MHz band is taken by the audio signal, a suppressed sideband, and an audio guard band. Color information for the picture is carried in a subcarrier, rather than as an integral part of the video (as it is in the PAL and SECAM systems), because the FCC decreed that *color* television must be compatible with and be able to be seen (in black and white) by those sets which could receive only black and white. To have done otherwise would have immediately made obsolete all of the black-and-white television sets in use at that time, as soon as the networks began colorcasting.

The *color* video picture in the NTSC system can be viewed on a color video monitor or on a color television receiver, and at the same time the same picture can be viewed in black and white on a black-and-white monitor or receiver. (See Fig. 3–4A and B.)

A video monitor, whether black and white or color, differs from a receiver in that it has no channel tuner and so must receive its video signal (with or without audio) by a coaxial cable feed from a program source.

The Wave-Form Monitor and the Vectorscope

The sync signals and the electronic waveform of the picture can be viewed by studio technicians on a waveform monitor.

Figure 3–4A Color monitor. *(Provided courtesy of Sony.)*

Figure 3–4B Waveform monitor and vectorscope. *(Provided courtesy of Tektronix.)*

The home viewer, of course, would have no need for this specialized information. The waveform monitor is a specialized oscilloscope which looks like a picture monitor externally, but has a monochromatic black-and-green screen (Fig. 3–5). The screen depicts a waveform display in graphic form of video amplitude and synchronizing signals versus a time base.

The waveform monitor is used primarily by the video technician *during* a program to "ride" video levels and to check the sync relationships displayed on the graticule of the monitor. The *graticule* is an overlay on the monitor screen that displays the correct relationships of the

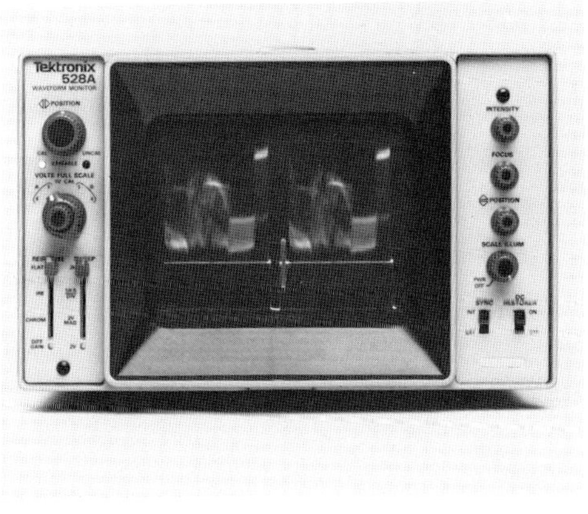

Figure 3–5 Waveform Monitor Display of two lines of video. *(Provided courtesy of Tektronix.)*

waveform, in terms of time (in microseconds) and level (in IRE units, which, for our purposes, are arbitrary measurement units).

The waveform monitor is used during nonprogram time in the control room by technicians to analyze the camera and camera system functions and to measure the overall amplitude of that control room's video system, known as *insertion gain,* which should be 1 volt (V) peak to peak. If the system amplitude is either more or less than 1 V, then that is an indication of distortion in the signal chain which must be traced to its source and corrected.

Our concern here, however, is the use of the waveform monitor during a program. Look at Fig. 3–6.

HORIZONTAL AND VERTICAL INTERVALS

We check to see that the ratio of picture amplitude to sync amplitude is maintained at 100 IRE units of picture to 40 IRE units of sync. The color burst amplitude is to be equal to sync amplitude and centered on the blanking level, the *back porch.* There must be a minimum of 8 cycles of color burst. Picture black must be separated from blanking by 9 IRE units. The timing of the TV signal must con-

Field Blanking Interval, NTSC

Code	Interval	Time
V	Field Period (ms)	16.667
H	Line Period (μs)	63.5
J	Field Blanking Period	(19.5-21)11 +11.1 μs
K	Rise Time of Field Blanking Edges (10-90%)	6.35 μs
L	Duration of First Equalizing Pulse Sequence	3H
M	Duration of Sync Pulse Sequence	3H
N	Duration of Second Equalizing Pulse Sequence	3H
P	Duration of Equalizing Pulses	2.54 ±.1 μs
Q	Duration of Field Sync Pulse	27.1 μs
R	Interval Between Field Sync Pulses	4.44 ±.45 μs
S	Build-up Time (10-90%) of Sync Signal Edges	0.25 μs

Line Blanking Interval, NTSC

Interval Period	Time
Line Period	63.5 μs
Line Blanking Interval	11.1 ±0.2 μs
Front Porch	1.59 ±0.05 μs
Sync Pulse	4.76 ±0.2 μs
Rise Time (10-90%) Blanking to peak white	0.64 μs
Rise Time (10-90%) (Line sync)	0.25 μs

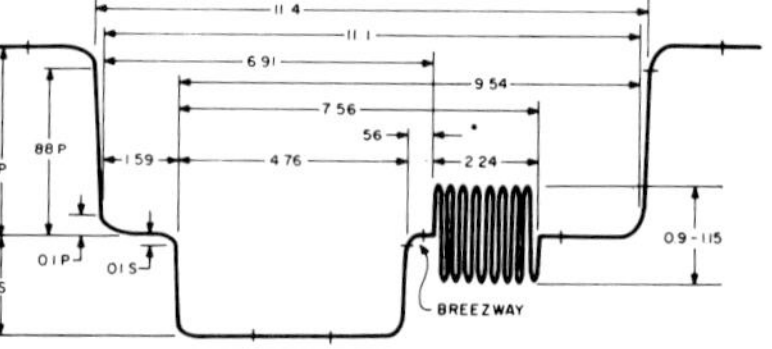

Figure 3–6 Pictorial representation of NTSC field and line blanking intervals. *(Provided courtesy of RCA.)*

form to a format diagramed in FCC specification no. 73.699. During the vertical interval, vertical blanking must start three lines before the start of the vertical sync pulse, and the relative widths of the pulses must be approximately correct.

Horizontal blanking must be between 10.49 and 11.44 microseconds in width. The horizontal sync pulse must be between 4.45 and 5.08 microseconds in width.

The *front porch*, that area between blanking and the leading edge of horizontal sync, must be no less than 1.27 microseconds in width.

The period between the trailing edge of the horizontal sync pulse and the first cycle of color burst, known as the *breezeway*, must be no less than 381 nanoseconds, or 0.381 microseconds. The time between the end of the color burst and the leading edge of the next line of picture is not specified since its timing is defined by the other parameters and by blanking width requirements. The frequency of the color subcarrier or color burst signal must be held within 10 Hz of 3.579545 MHz (usually abbreviated 3.58 MHz).

The rise and fall times of the horizontal sync pulse, measured between the 10 and the 90 percent points on the leading and trailing edges of the pulse, must be less than 0.250 microsecond (250 nanoseconds).

The 5-microsecond back porch is between the horizontal sync pulse and the blanking pulse. (See Fig. 3–7.)

Vertical blanking is the time between the last picture information at the bottom of one field and the first picture information at the top of the following field. Vertical blanking is measured from the leading edge of the first equalizing pulse. In terms of time, vertical blanking must

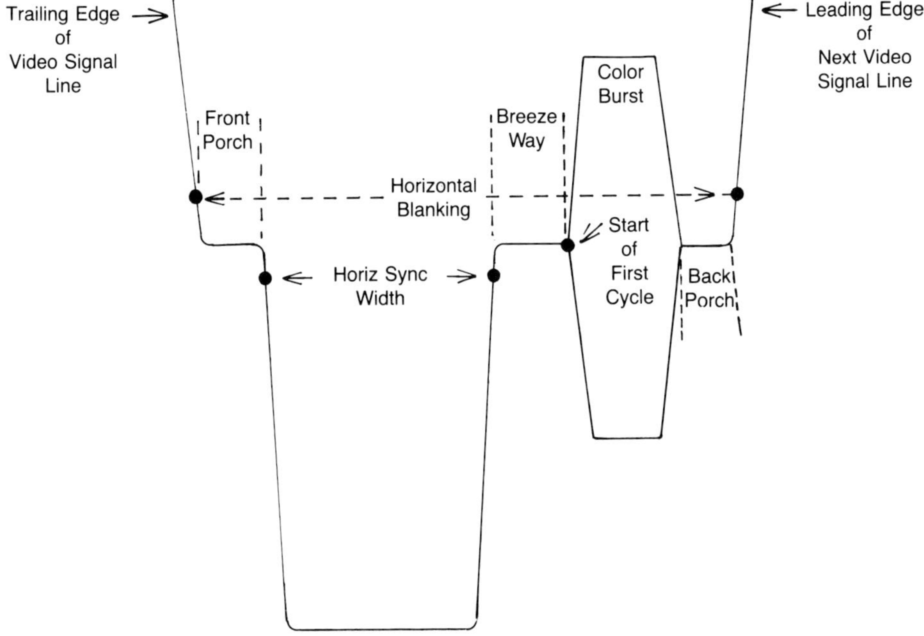

Figure 3–7 Pictorial representation of horizontal blanking pulse, horizontal sync, and color burst.

be greater than 1.17 milliseconds but less than 1.33 milliseconds. In terms of scanning lines, the maximum vertical blanking is 21 lines. It is fairly standard in the industry to stay with 21 lines of blanking, because the vertical-interval lines preceding the picture are usually used for transmission of vertical-interval test signals.

The width of the equalizing pulses which precede and follow the vertical sync pulse should be 2.54 microseconds. The tolerance permitted for equalizing pulses is that the area of an equalizing pulse must be between 45 and 50 percent of the area of a sync pulse.

The vertical sync pulse should have a total width equal to three horizontal scanning lines. The serrations in the sync pulse must be between 3.8 and 5.1 microseconds, measured at the −4 IRE level. (See Fig. 3–8.)

All the technical information just given describes the information found *between* the lines of picture and is easily identified by a marked graticule on the waveform monitor.

The waveform of the *picture line* itself describes the amplitude of the picture line, in both positive and negative directions (peak to peak), above and below a baseline. And the higher the line (in the positive direction), the lighter the shade of picture, or color of fleshtones, up to 100 percent white, or maximum picture brightness.

The waveform also depicts the opposite condition to maximum brightness (in the negative direction), or *color black*, which is the darkest part of the picture. This area is also called *pedestal*, or *picture black*. Pedestal describes the lowest picture voltage in the visible picture.

All the sync pulse information is indicated on the waveform graticule as being

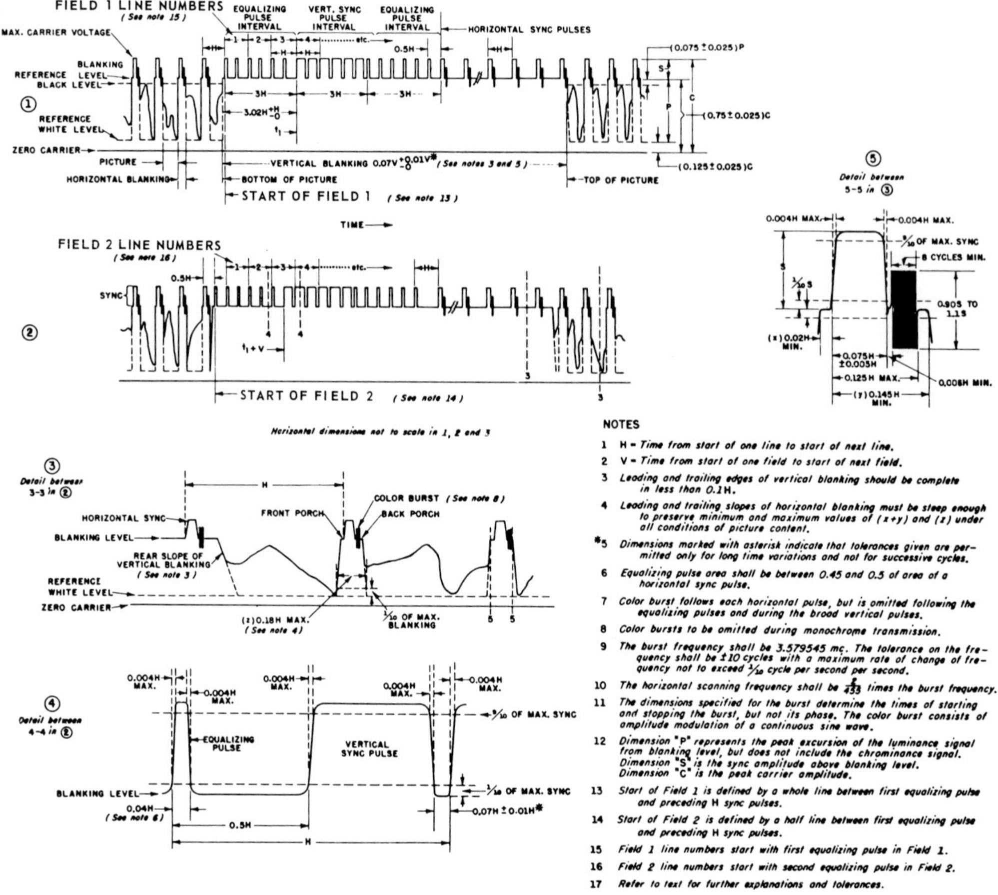

Figure 3–8 Diagram. *(Provided courtesy of Tektronix.)*

located in a region on the waveform *below* and separated from pedestal, in a region known as "blacker than black."

The vectorscope is an oscilloscope which is calibrated to display the color *phase* relationships between any two signal sources, where the sources are the red, green, and blue chrominance of the video picture and the color burst frequency is 3.58 MHz. (See Fig. 3–9.) The vectorscope is used by the control room video technician to maintain color relationships, so that various color picture sources can be color-matched, and to make the color that is displayed on a viewer's receiver as true

a reproduction of the original as possible. Recall that the color information in a television signal modulates the 3.58-MHz color subcarrier (the color burst), to lock in the color information. In the receiver, or color monitor, this signal triggers a 3.58-MHz oscillator which is timed and phased to the camera color signals and which demodulates the encoded color information.

Every camera, camera control unit (CCU), sync generator, video tape recorder (VTR), and switcher will have available a source of color bars test signal, as will every recorded tape, just preceding

Figure 3–9 The vectorscope. *(Provided courtesy of Tektronix.)*

program. And each of these devices will have controls available to *shift the phase* of the color burst by individually shifting the phase of the red *(R)*, green *(G)*, and blue *(B)*. The calibrated graticule of the vectorscope screen shows the positions of the color burst and the individual color signals generated by the color bars test signal. Before recording or playback of a program, all program sources are aligned by using the color bars signal, so that the position of the burst and the phase of each of the *R*, *G*, and *B* signals generated by the bars bear a fixed relationship to each other, as indicated on the vectorscope graticule. Where these relationships are found to be off, the *R*, *G*, and *B* controls are adjusted.

What we have, then, is a video picture conceived in the camera, with sync and color timing information added to the picture by an external sync generator, or by one built into the camera, to combine in a mix called the *composite video*, or *composite signal*. This is defined as a signal containing all of the necessary information needed to reproduce the video picture. A *synchronous signal* is one in which *all* sync pulses are in time coincidence with the reference signal. A *phased signal* is one whose color burst is running in coincidence with the reference signal. And a *reference signal* is the one selected signal against which all other signals are judged.

If only one camera is to be used for a program, its composite signal can directly feed an air or cable channel or into a VTR.

However, often several cameras as well as other sources of video, such as VTR playbacks, special-effects generators, text or graphics generators, and telecines (video from film or slide projectors), are integrated to make a video program. This mix requires that each video source be fed into a mixing device called a *production switcher* and that all of the inputs to that switcher be synchronized with a reference signal from one sync generator or with a *genlock* circuit, which locks all inputs to a single source of synchronization.

More on Video Cameras

Before we take a look at a few brand-name cameras and camera systems, with particular emphasis on how they compare regarding specific features, let us go back to camera basics and pick up some information that we deliberately omitted until now for clarity of presentation.

In the cameras of early black-and-white television, there was only one monochromatic pickup tube. The tube was quite large, as was the camera. It took two strong people to lift the camera. When the NTSC color system emerged, it was necessary to redesign the camera with a three-tube system in which the picture information is separated into *luminance,* or black-and-white picture information, and *chrominance,* or color information. The luminance information is gathered by the color green pickup tube, which becomes a sort of baseline from which other colors gathered by the color red and color blue tubes are added or subtracted to provide the entire color spectrum seen in color video.

With three pickup tubes, the lens system must feed light into a *camera optics system,* which, by using mirrors and/or prisms, divides the incoming light and separates it for the *R, G,* and *B* tubes. Further, there is the color temperature-controlling filter wheel or disk. It is located adjacent to the lens and is thumb-controlled by the camera operator. This wheel or disk has a closed, or *capped,* position, allowing no light to enter the camera; a 3200-degree Kelvin position for indoor incandescent lighting; and a 5600-degree Kelvin position for outdoor sun lighting. Most often there is an additional 5600-degree Kelvin filter position which includes a neutral-density filter to reduce the light intensity and the color temperature of sunlight to the 3200-degree Kelvin range over which the camera is designed to operate.

The lens mount is manufacturer-designated. It is either the screw-in C mount or the *bayonet,* push-and-twist mount.

There is a camera video amplifier *gain select* switch on the camera, which can be set for 0, +6 decibels (db), or +12 db, or something similar like 0, +9, and +18 db. This switchable gain amplifier is normally left at the zero level setting. It can be set to the higher settings to compensate for low light conditions during camera operation, by raising the gain of the amplifier. Use of this amplifier is normally avoided, because as the camera video gain is raised, the *grain* (the noise) in the picture is also increased.

Most cameras have *tally* lights, operated automatically by the switcher, that indicate, when illuminated, that the camera is on. One tally light is on top of the camera for the information of the people in the studio. Another, a small light-emitting diode (LED) indicator, is located in the camera viewfinder hood for the camera operator.

There is a *white-balance* circuit and a separate *black-balance* circuit, either automatic or manual, often with a battery-operated memory chip, or integrated circuit, that remembers the settings even after the camera power has been turned off. White balance and black balance are camera operating baselines, and they must be carefully maintained. Whenever the color temperature of the light source has changed, from indoor to outdoor lighting or from incandescent to fluorescent light, the white balance of the camera *must* be reset.

White balance can be reset or corrected by aiming the camera at a pure white source and momentarily actuating its circuit with a switch.

There is circuitry, controlled by a switch on the camera, that provides either camera video or color bars signal at the camera output. Thus one camera can be compared to another by using the color bars signal within a studio.

These and other setup procedures are discussed in detail later.

Camera manufacturers speak of a video camera in their sales literature in segments: the camera head; the lens system which accompanies the camera or optional lenses, most often manufactured by another company; and the pickup tubes, often a choice of Plumbicons, Saticons, or Trinicons or others, manufactured by still another company. Then there is the camera control unit, the varying power supplies, and the camera mounting devices.

Here, simply to get a sense of the cameras available and to detail their features, we describe some of the various brand-name cameras in terms of their operating features. We include details from the manufacturer's specification sheets and provide photographs and diagrams, where applicable, of the operational parts of the camera or camera system. Since this text was conceived and written specifically within the context of cable television, we limit the greatest part of our discussion of cameras to those costing approximately $7000 to $10,000 at this writing. We do, however, include a couple of cameras both above and below this range for comparison. Video cameras are price-rated as high-end or low-end broadcast models, high-end or low-end industrial models, and consumer electronics models. Often there is a very thin line separating the quality and price of low-end broadcast from high-end industrial models.

The cameras discussed here are some of the camera types currently in use at many cable TV installations and some of the educational institutions reviewed by this author. These cameras are of much higher quality and are much more expensive than home consumer products, but are considerably less expensive and of lower quality (in terms of resolution, lines per inch, and color clarity) than broadcast studio cameras found in the $35,000 to $55,000 price range. The cameras shown here, in addition, can do double duty, both in the studio and in the field, with the exception of the RCA broadcast camera.

Note that inclusion or exclusion of a piece of television equipment in this discussion does not imply either endorsement or dislike of that equipment.

The cameras all have two viewfinders available. One has a viewing diameter of about 1½ inches and an eyepiece for close viewing, normally under out-of-studio conditions. The other is a studio viewfinder with a diameter of about 4½ inches. Both viewfinders are, in reality, small black-and-white video monitors which are mounted on, but not part of, the camera and are fed, either from the camera output or by return feed, back through the camera cable for system monitoring.

JVC KY 2700AU Three-Tube Color Video Camera

The JVC (Japanese Victor Corporation) KY 2700AU camera has three ⅔-inch Saticon tubes with electrostatic focusing and electromagnetic deflection, for red, green, and blue. (Saticon is a registered trademark of Hitachi.) It has a parallel-mirrors optics system in a die-cast aluminum camera head. It weighs 12.6 pounds without the lens or viewfinder. The lens systems are C (threaded) mounted, and the printed-circuit boards are the plug-in type. The white-balance circuit is automatic, with memory. (See Fig. 3–10.)

Iris control is automatic (or manual if the cameraoperator wishes), as is black balance, and there is a built-in genlock circuit and color bars generator.

Remote control of adjustments is available when the camera is used in conjunction with the RS 2000 camera control unit.

Figure 3–10 The JVC KY 2700AU photo—left side. *(Provided courtesy of JVC.)*

The RS 2000 provides camera power through the camera cable for up to 980 feet of cable.

It has a signal-to-noise *(S/N)* ratio of more than 54 db, a sensitivity of $f/4.0$ at 2500 lux (250 footcandles), and 3200-degrees Kelvin color temperature. Its minimum illumination at $f/1.6$ is 100 lux (10 footcandles) with the $+12$-db switch on, and there is a resolution factor of 600 lines at the center of the green channel. There is automatic beam control which the manufacturer states will eliminate comet tailing from moving objects. *Comet tailing* is a swirling line or streak which is inadvertently created electronically in the picture when a predominantly shiny object moves rapidly, such as the protective helmet on a baseball batter or football player, in bright sunlight. The camera requires power of 12 V direct current (dc) at 1.45 amperes without the viewfinder. It can operate within an ambient temperature range of 23 to 113°F.

The KY 2700AU uses one of two available servo-powered (motor-driven) zoom lens systems.

HZ-2140U Zoom Lens. This is the lens recommended by JVC for the KY 2700AU because it ensures correct operation of the automatic iris control mechanism. This lens has a zoom ratio of 14:1 with focal lengths from 10 to 140 millimeters (mm). Its effective aperture is $f/1.7$ from 10 to 100 mm and $f/2.4$ out to 140 mm. The zoom mechanism can be driven by the servomotor, or it can be zoomed manually. The iris control is also switchable betwen automatic and manual. (See Fig. 3–11.)

The HZ 2100U Zoom Lens. This servo zoom lens has a zoom ratio of 10:1 with focal lengths from 10 to 100 mm. Its effective aperture is $f/1.6$. It also has macro-focusing capability, which permits a lens-to-subject distance of as little as 0.5 inch minimum for extremely tight closeups or for zooming in on very small objects. Both zoom and iris control can be done manually or automatically. (See Fig. 3–12.)

Two electronic viewfinders are available for this camera: a 1.5-inch viewfinder with eyepiece and a 5-inch viewfinder for studio work.

The 1.5-inch viewfinder weighs 1.8 pounds and is attached when the camera is used as a portable (Fig. 3–13). The finder

Figure 3–11 HZ 2140U zoom lens. *(Provided courtesy of JVC.)*

Figure 3–12 HZ 2100U zoom lens.
(Provided courtesy of JVC.)

can be moved back and forth away from the camera about an inch, and it can be tilted up and down for ease in viewing when the camera is held above or below eye level. Red and green LEDs in the eyepiece indicate when VTR recording is occurring and give low-battery warnings for

Figure 3–13 VF 1900BU viewfinder.
(Provided courtesy of JVC.)

both the camera and a VTR as well as a warning of tape end or incorrect tape winding. These functions vary depending on the VTR being used. The viewfinder has a tally and white-balance LED indicator, and a video signal level waveform can be switched to appear on the finder screen, superimposed over the picture.

The 5-inch viewfinder is mounted to the camera for studio operation (Fig. 3–14). It can be tilted 30° up or down and can be locked in any position within this range. It can utilize the camera-supplied video signal level indicator and white-balance indicator on its screen as well as tally in the viewfinder hood. When the camera is set up for studio applications, several flexible-cable drive units are available that mount on the pan-tilt handle and remotely operate the zoom and focus controls.

RS 1900 Remote Camera Control Unit. As we inferred earlier, the JVC KY 2700AU camera can be used in a number of system applications, and these will be described; but whenever the camera is not being used as a portable, it is used in con-

Figure 3–14 VF 2500BU viewfinder.
(Provided courtesy of JVC.)

junction with a camera control unit, the RS 1900. (See Fig. 3–15.)

By using the control unit, the camera adjustment functions can be controlled from up to 300 meters (980 feet) of cable away. Then the camera operator is free to devote all his energies to composing the pictures and being artistically creative. Meanwhile, a technician in the control room can use the RS 1900 to operate the camera remote controls, including power on/off, automatic or manual iris control mode selection, color bars or camera signal selection, gain selection, automatic white-balance control, red or blue horizontal and vertical shift, master black level control, and red or blue gain control.

The RS 1900 supplies camera power over the same cable that feeds the picture from the camera. Camera cables are available in a number of lengths which require length compensation, and the RS 1900 can compensate for cable length with both coarse and fine adjustments. Other RS 1900 features are its genlock and auxiliary inputs, horizontal phase and subcarrier phase controls, and separate red, green, and blue outputs. Intercom and tally facilities are also built into the RS 1900.

AA-C20U AC Power Adapter. When the camera is used without the camera control unit, that is, it feeds a VTR directly, it is powered by either a battery supply or the power adapter. The AA-C20U can be attached directly onto the back of the camera in place of the battery pack. Or if weight is a consideration, it can be connected to the camera via a power supply cable.

Figure 3–16 describes some of the system applications of the JVC KY 2700AU camera.

Hitachi FP-21 and FP-22 Cameras

Made by Hitachi Denshi America, the FP-21 is a three-tube camera employing ⅔-inch Saticon tubes. The FP-22 is the same camera in all respects except that it has an additional printed-circuit board featuring a microprocessor that digitally controls an automatic setup system, including black balance, white balance, centering, pulse canceling, and fault diagnosis functions. (See Fig. 3–17.)

If, during the automatic setup or the operation of the FP-22, any faults are discovered electronically in the circuitry controlling the white or black balance, the centering, or the pulse canceling functions, then an LED indicator lights inside the camera viewfinder, and identifying LEDs light on the faulty printed-circuit board(s) to indicate the source of trouble.

The camera has prism optics, a low-noise preamplifier which provides a res-

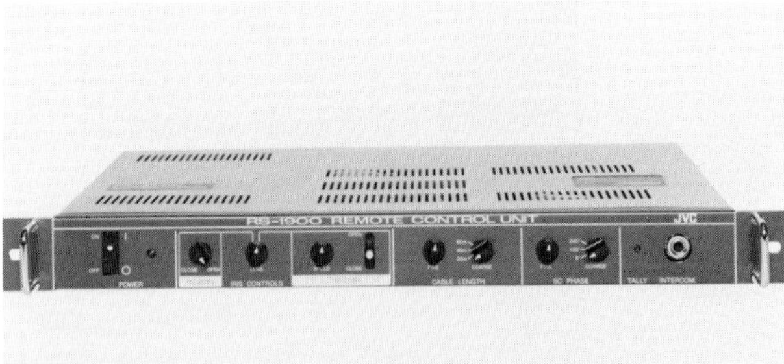

Figure 3–15 RS 1900, front view, showing controls. *(Provided courtesy of JVC.)*

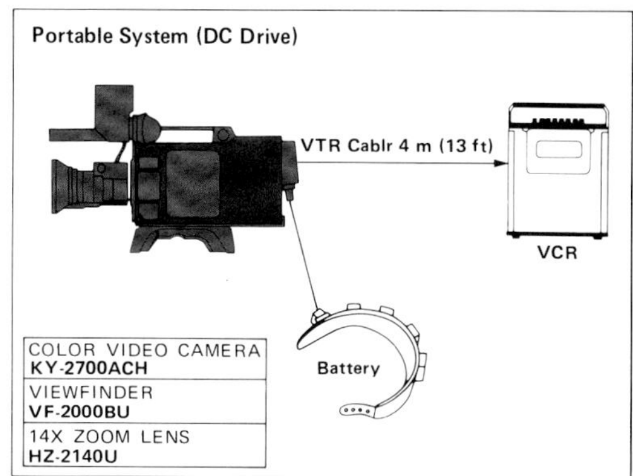

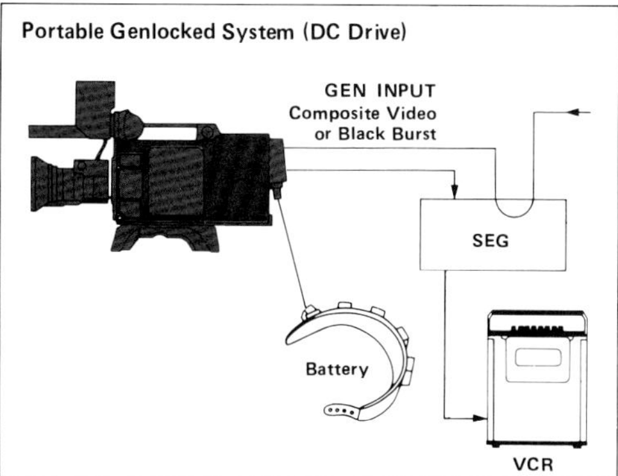

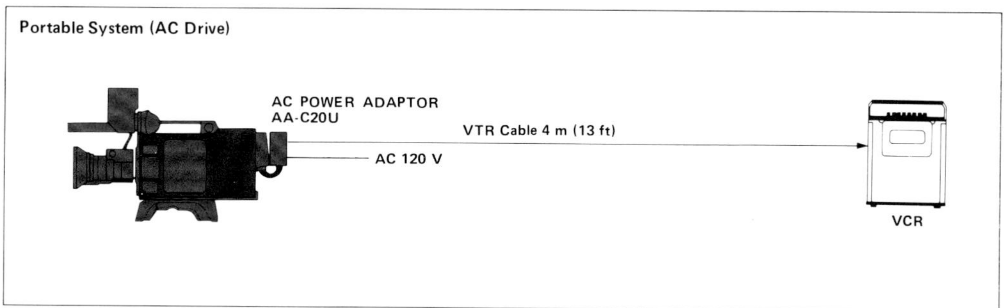

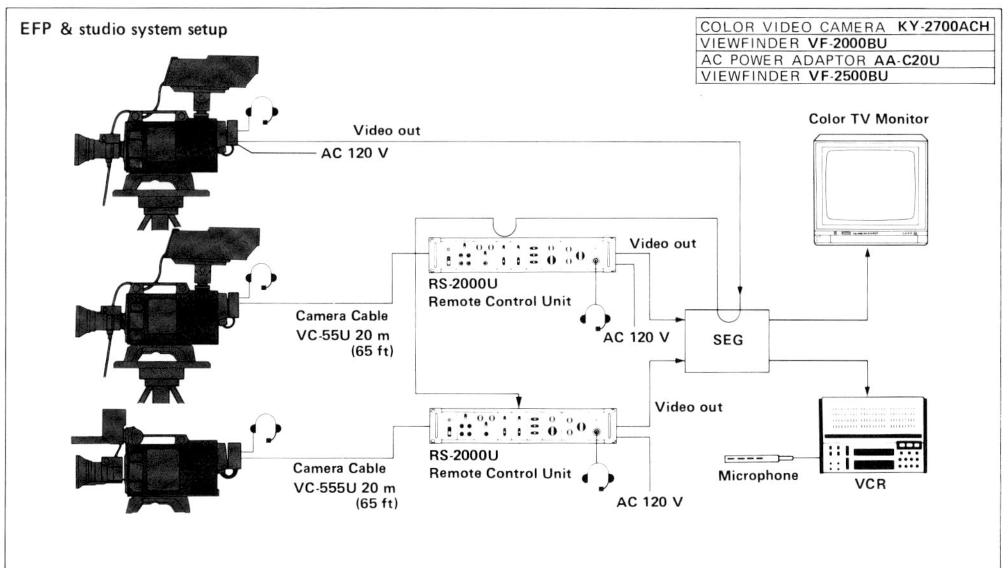

Figure 3–16A Portable system (DC power), portable genlocked system, portable system (AC power), EFP and studio system. *(Provided courtesy of JVC.)*

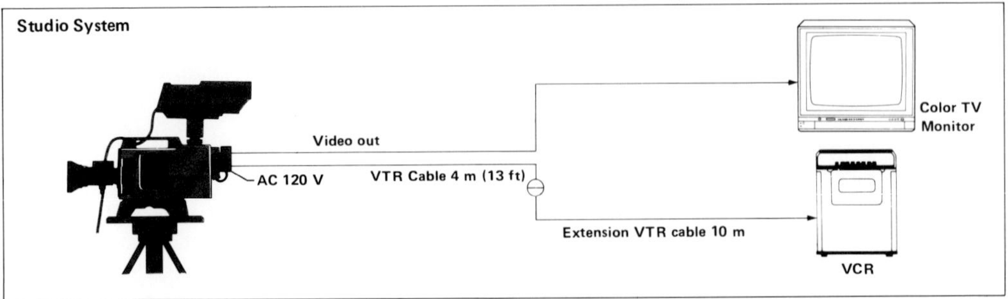

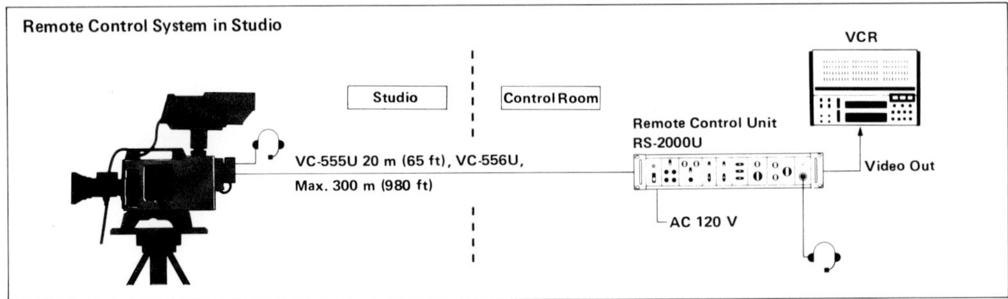

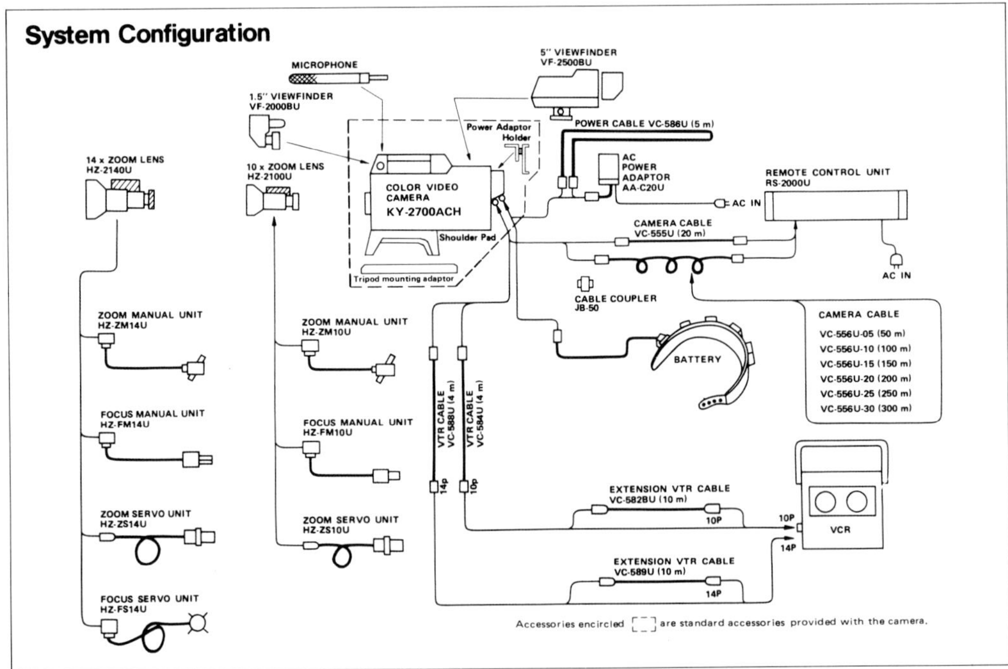

Figure 3–16B Studio system, remote control system, and total system configuration. *(Provided courtesy of JVC.)*

Figure 3–17 The Hitachi FP21/FP-22 camera. *(Provided courtesy of Hitachi.)*

olution factor of 580 lines at the center of the green channel, and a signal-to-noise ratio of 55 db.

Its sensitivity at $f/4.0$ is 2000 lux (200 footcandles), and its ambient operating temperature is rated from -4 to $+104°F$. It requires 12 V dc operating power.

The camera head is constructed of a magnesium alloy casting, and the camera weighs about 11.5 pounds without lens or viewfinder. The lens mount is the bayonet type.

White balance is attained in 0.5 second with the automatic white switch flipped up and the camera aimed at a pure white object; black balance is attained when the automatic black switch is flipped down.

The automatic iris is triggered by the video signal level. By opening and closing the iris to some degree during wide variations of light conditions, a constant video signal level is maintained. Control of the iris is achieved by detection of the peak values of red, green, and blue signals, rather than green alone, thus providing natural color control of the iris, even when the object is almost monochromatic. The automatic iris closes during the following situations: when the FP-22 is set to the automatic black and automatic pulse-cancel modes of the automatic level set sequence; for about 4 seconds after the automatic black switch has been thrown; when the camera is set to the color bars or the standby power mode; and when the camera power switch is turned off, which protects the camera tube targets.

The camera gain switch provides 0, +9, or +18 db of picture amplifier gain.

The power switch operates in on or standby mode. Returning to on from standby requires 10 seconds for the tube filaments to reheat. The optical filter disk has four positions: 3200 degrees Kelvin, 5600 degrees Kelvin, 5600 degrees Kelvin + ⅛ ND (neutral density), and CAP. This filter, together with the automatic iris and automatic white balance, allows the FP-21 to cope with wide variations in light level and color temperature.

Registration adjustment, which entails the horizontal and vertical centering of the red and blue channels, can be performed on the camera by using its controls or the controls of the camera control unit, the ROU.

Two lens configurations are available for this camera: a 10:1 zoom ratio with focal length from 11 to 110 mm and a 14:1 zoom ratio lens. These are servo-controlled lens systems, and a lens cable kit provides servo or manual control of focus and zoom.

Two viewfinders are available: the GM-3B, which is a 1.5-inch viewfinder with a diopter correction eyepiece for camera operators who normally wear glasses, and the GM-5B. The GM-3B can be moved to the left and to the right for optimum positioning. A waveform display of video level is available for switch selection, superimposed on the picture, for checking the video level. There are also LED indicators for white balance, tally and VTR recording, diagnosed fault alarm on the FP-22, and low-battery alarm provided inside the 1.5-inch viewfinder.

The GM-5B viewfinder is a studio 5-inch finder which mounts on the top of the camera and can be adjusted to any viewing angle for maximum visibility.

The ROU OP 22A/21A is the camera control unit that is used with the Hitachi camera when it is part of a system in a control room or studio setting or when the system is used for out-of-studio electronic field production (EFP).

The camera adjustment and control functions that may be assumed by the ROU are as follows: iris, master black, *R* and *B* black level, *R* and *B* gain level, *R* and *B* horizontal and vertical centering, high gain select, camera or color bars select, power on/off, automatic white and automatic black balance, automatic centering and automatic setup on the OP 22A, subcarrier phase, horizontal phase, camera call, tally, and cable length compensation.

The power source needed by the camera is 12 V dc, which can be obtained from the ROU, a battery belt pack, an ac adapter, or an automobile battery. When the battery voltage drops below the required voltage, an alarm LED inside the viewfinder alerts the camera operator.

In addition to the video line out and VTR outputs, there is separate monitor output. The external signal can be fed as a return video input to the viewfinder for monitoring, and a built-in sync generator can be genlocked to an external composite or black burst signal. The horizontal blanking width can be adjusted from 10.5 to 11.5 microseconds and the vertical blanking width from 18 to 21 lines. This makes it possible to compensate, in advance, for the errors introduced into the

recorded video signal in the recording/time-base correction/editing process.

Figure 3–18 describes the operating systems applications.

Sony DXC 6000 Color Video Camera System

The DXC 6000 is a three-tube ⅔-inch Saticon camera with prism optics. The camera head is made of aluminum, and it uses a bayonet lens mount. (See Fig. 3–19.)

The gain increase available is +9 and +18 db, and there is a signal-to-noise ratio of 53 db. The camera has vertical and horizontal image-enhancing circuits, and both the camera head and the camera control unit (CCU) have microprocessors that handle many setup functions and monitor systems operations. Serial-data transmission between the two units reduces the number of conductors needed in the camera cable, allowing a smaller-diameter cable; and the head memory retains setup instructions during power interruption or failure. Automatic controls include a beam optimizer to minimize comet tailing, iris, digital black-and-white balance, with lens iris closing, and battery-powered memory to preserve black-and-white balance during power shutoff.

The camera has a genlock circuit, audio and video monitoring for recording and playback, test signal generator, color bars generator, adjustable blanking width, flare and shading compensation, return video, power standby, level indication, and test output. From the CCU (Fig. 3–20) beam set and wobbling switching is available.

The CCU is available for studio control of the camera or for EFP work (Fig. 3–20). Four Fujinon lens systems and a 1.5- or 4-inch viewfinder are available. Power is

OPERATING SYSTEMS

Ample accessories offer you various systems configuration serving your purpose.

(1) ENG application: Operating the camera with batteries in combination with a portable VTR.

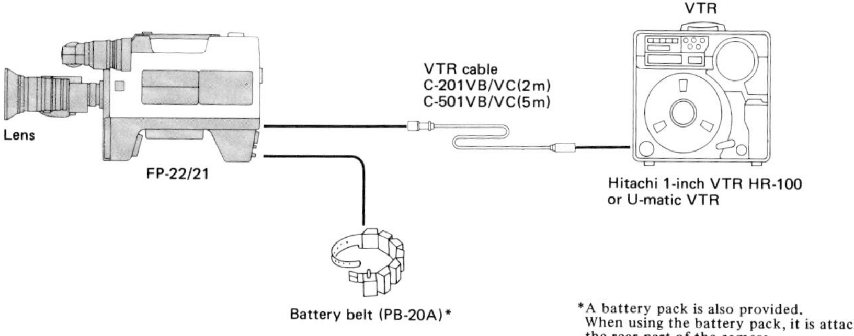

(2) ENG application: Operating the camera on AC in combination with a portable VTR.

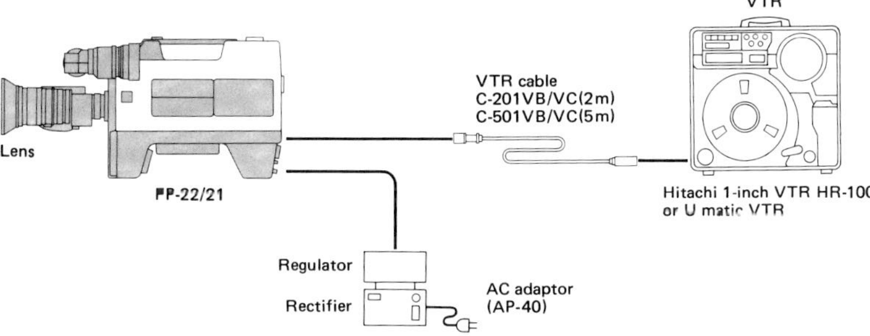

(3) EFP application: Using the camera for EFP in combination with ROU.

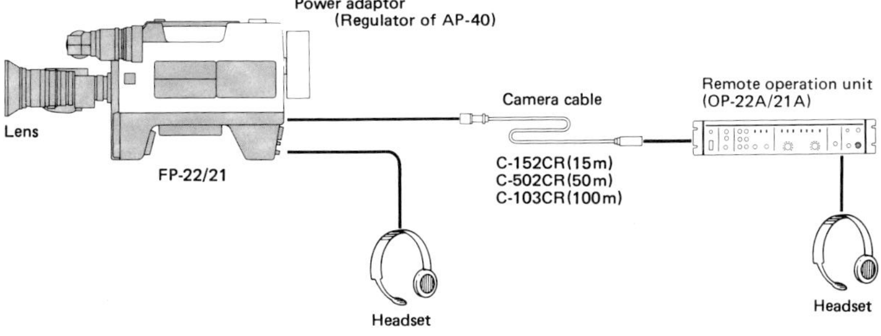

Figure 3–18A ENG and EFP applications of FP-22 Hitachi Denshi. *(Provided courtesy of Hitachi.)*

(4) Studio application: Using the camera as a studio camera in combination with ROU.

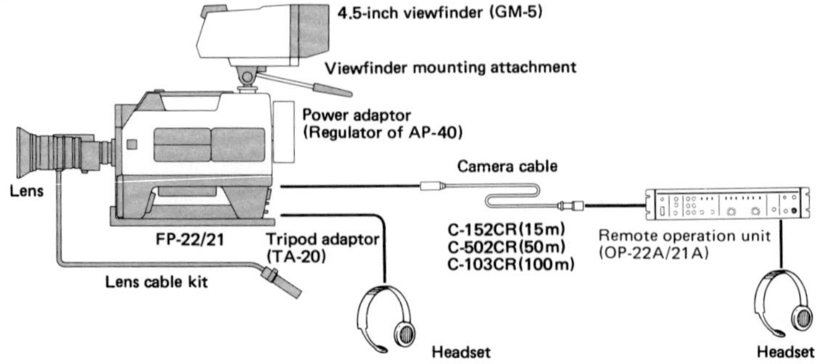

Figure 3–18B Studio application of FP21/22 Hitachi Denshi. *(Provided courtesy of Hitachi.)*

Figure 3–19 The Sony DXC 6000 side view. *(Provided courtesy of Sony.)*

Figure 3–20 The CCU-6000. *(Provided courtesy of Sony.)*

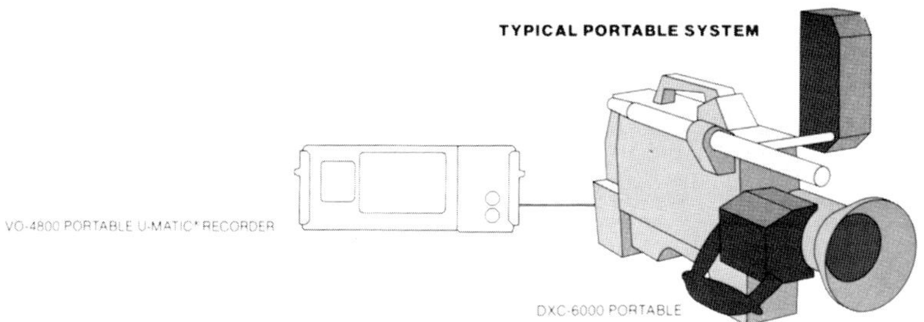

Figure 3–21A Typical portable system diagram. *(Provided courtesy of Sony.)*

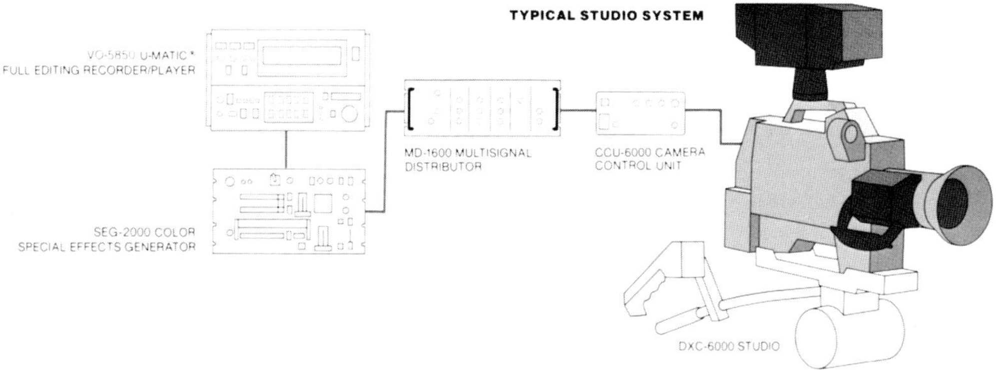

Figure 3–21B Typical studio system diagram. *(Provided courtesy of Sony.)*

supplied by a battery pack or by an ac adapter.

Ikegami ITC-730 Video Camera

The Ikegami ITC-730 is a three-tube ⅔-inch Saticon camera with prism optics. The head is of die-cast magnesium alloy, and it has a bayonet lens mount which permits a wide range of Fujinon lenses to be used. (See Fig. 3–22.)

The camera has automatic iris which permits light control from peak levels to average. The automatic iris closes automatically to prevent damage to the pickup tubes upon loss of power. The color temperature filter wheel has settings for 3200°K, 5600°K, 5600°K + a 25 percent neutral-density filter, and a capped position. The ITC-730 uses digital white balance which is achieved instantly and retained in memory for over a year with the camera unpowered. The +6/+12 db video gain switch makes it possible to double or quadruple the camera sensitivity, enabling shooting at lighting conditions as low as 60 lux with an f/1.4 lens.

A "zebra pattern" light level indicator is displayed in the viewfinder to optimize light level adjustment, and the built-in color bars generator conveniently matches color between cameras as well as adjusts color monitors. The zebra-pattern circuit creates a diagonal zebra stripe pattern over the video picture in the viewfinder when the light level is too high and a flashing pattern when the light level is too low.

Figure 3–22 The Ikegami ITC-730 camera. *(Provided courtesy of Ikegami.)*

Return video can be viewed on the viewfinder. Power consumption can be reduced to 1 W during standby, by using the standby switch.

The eyepiece on the 1.5-inch viewfinder is adjustable in position, angle, and vision correction. Tally lamps indicate "on-air," "VTR-start-stop," and cameraman call from the CCU.

The CCU-730 remotely controls a series of functions that permit the camera operator in the studio to give complete attention to pictures (Fig. 3–23). For studio use a 5-inch viewfinder is available, and the camera can be powered by a battery pack, by an ac adapter to the camera, or through the CCU-730 camera control unit via the camera cable. The CCU-730 supplies a fixed 12 V to the camera re-gardless of variation in the length of the camera cable. (See Fig. 2–24.)

We have examined four of the many three-tube cameras that are on the market for use on cable television. Often these cameras are also found at educational institutions which teach television operations. Now we examine one single-tube broadcast-quality camera and then a broadcast studio camera.

Sony BVP-110 Camera

The recent marketing by Sony of the HBST (highband Saticon Trinicon) pickup tube has made possible a high-quality single-tube color camera. The Sony BVP-110

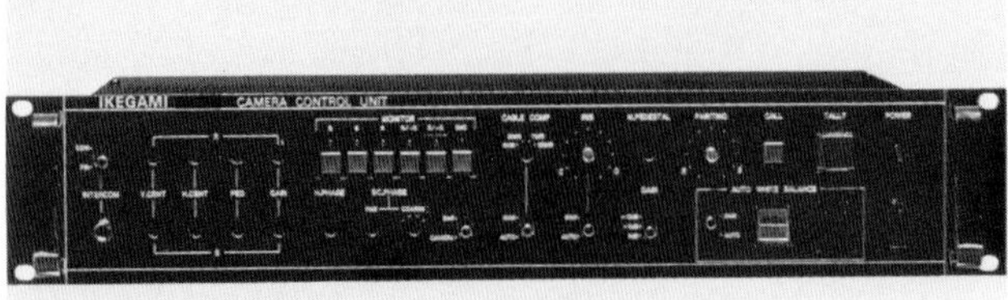

Figure 3–23 The CCU 730 camera control unit. *(Provided courtesy of Ikegami.)*

	Hitachi FP-21/22	JVC KY-2700	SONY DCX 6000	Ikegami ITV 730
Pickup Tube[a]	2/3 in. Sat	2/3 in. Sat	2/3 in. Sat	2/3 in. Sat
Optical System	Prism	Mirror	Prism	Prism
Horiz. Resolution[b]	580	600	500	600
Sensitivity[c]	2000	2500	2000	2000
Min. Illumination[d]	40@f/1.6	100@f/1.6	40@f/1.6	60@f/1.4
Registration[e]	.1,.2,.4	.1,.2,.4	.1,.4,.8	.1,.2,.4
High Gain	+9,+18db	+6,+12db	+9,+18db	+6,+12db
Filters[f]	32,56,	32,45,	32,56,	32,56,
	56+1/8	60+¼	56+¼	56+¼
Video Out Level	1V@75	1V@75	1V@75	1V@75
S/N	55 db	54 db	53 db	54 db
Operating Volts	12VDC	12VDC	12VDC	12VDC
Power Used	22W	17.4W	21W	18W
Weight	11.5 lbs	12.6 lbs	12.3 lbs	9.9 lbs

[a]Sat = Saticon.
[b]Lines, in green channel.
[c]Lux, at f/4, 3200° Kelvin.
[d]Lux, at 3200° Kelvin.
[e]Zones 1, 2 and 3, where 1st zone is circles, 80% pix height, 2nd zone is circle, of pix width, 3rd zone is area outside 2nd zone.
[f]Numbers are × 100, (I.e., 3200), in ° Kelvin, +% Neutral Density Filter.

Figure 3–24 Camera specifications comparison chart.

camera system is one such camera (Fig. 3–25).

The BVP-110 is a single-tube HBST camera which contains less than half the components and one-third of the alignment controls found in conventional three-tube cameras. It is rugged and stable and is a fully integrated camera package; that is, there are no external cables to connect (such as zoom lens powering or viewfinder cables), except for the cable(s) connecting the camera to a VTR or to a system.

It has an internal genlock sync pulse generator, a color bars generator, and an image enhancer. The automatic white-balance system employs separately powered memory and uses any white scene source for balance. A preset 3200-degree Kelvin white-balance source is also provided.

The BVP-110 weighs 6 pounds including the viewfinder but without the lens.

Figure 3–25 The Sony BVP 110 camera. *(Provided courtesy of Sony.)*

It operates on 11 W from a 12-V power source. There is a choice of two standard lens systems: 10×, f/1.6, 11 to 110 mm; and 14×, f/1.6, 10 to 140 mm. Both lenses have automatic iris and servo zoom.

The 1.5-inch viewfinder (there is no studio finder) contains a battery status indicator as well as indicators for white balance, iris, filter selection, low light, and video level. Return video from a VTR playback also may be selected at the viewfinder. There are remote VTR start and power-save mode selectors for control from the camera, in VTRs equipped with these functions.

Figure 3–26 describes the highband Saticon Trinicon tube. Because of the single tube, there are no prism or mirror optics, and the filter wheel has four positions: 3200° Kelvin, 5200° Kelvin, 5200° Kelvin + 0.25 ND filter, and 6800° Kelvin.

Camera resolution is claimed to be 400

lines, uniformly, again as the result of a single pickup tube. Sensitivity is 2000 lux at f/4. The luminance signal-to-noise ratio is 53 db, and minimum illumination is 80 lux, with an f/1.6 lens setting and a +12-db video gain setting.

Controls on the camera include power switching with preheat; standby and VTR; gain switching of 0, +6, or +12 db; output switching of camera or color bars; white balance in automatic, in memory, or in preset; VTR record mode select; and return video from the VTR. The viewfinder has controls for brightness, contrast, and tally on/off.

The BVP-110 output connectors are a BNC for 1-V peak-to-peak video at 75 ohms (Ω), a 14-pin VTR connector, and an audio monitor output.

The inputs are genlock and dc power.

Figure 3–26 HBST tube. *(Provided courtesy of Sony.)*

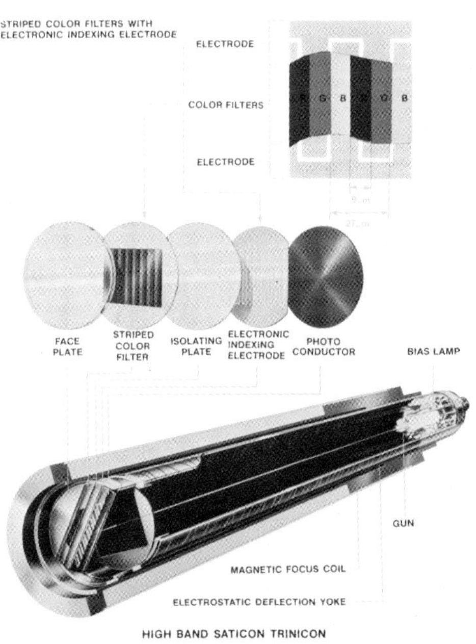

RCA TK-47B Studio Color Television Camera

The TK-47B (Fig. 3–27A and B) is an example of the level of equipment used at broadcast stations. Its system consists of three basic units: the camera head, the camera processing unit (CPU), and the remote-control unit (RCU). The CPU and the RCU need be connected only by twisted-pair cables. The CPU corresponds to the CCU on cable level equipment except that it has no operating controls. It functions as an interface device between the camera and the rest of the system. And since the CPU has no controls, it need not be located near the video control operator.

There are no operating controls on the camera, which frees the cameraoperator to perform only the creative part of the job. All of the camera control functions are performed in the control room on the

Figure 3–27A The RCA TK-47B camera. *(Provided courtesy of RCA.)*

Figure 3–27B The RCA TK-47B setup terminal. *(Provided courtesy of RCA.)*

RCU. There is one additional component to the system, the setup terminal (SUT), which is connected to the camera during setup and then disconnected during operation. A single SUT, therefore, can service all the cameras in a studio. There is also available an optional Autocam which reduces the setup operation to a single pushbutton for major camera setup adjustments. A "check" pushbutton verifies more than 60 functions in 15 seconds, using the channel green as a reference.

The TK-47B is a three-tube camera with prism optics. The optical system is shock-mounted, with the prism and tubes mounted in a single assembly on vibration isolators. A wide variety of lenses are available that use a quick-change mount.

There is a seven-position color temperature selector switch as well as a six-position color filter wheel. Six positions of master gain selection are available, and the lens cap automatically closes with power off. A two-level tally system is used. An LED display at the 7-inch viewfinder indicates whether the camera is on the air or being used for an "iso" (isolated camera) recording. The viewfinder is tiltable, rotatable, and removable. A sync generator with genlock is built in.

And finally, Hitachi Denshi and Sony have marketed—so far, for consumer product use only—a *no-tube* camera that uses, instead of a pickup tube, a *silicon chip sensor* no larger than a postage stamp. Thus the camera is much lighter, uses less power, and eliminates the prevailing problem caused by pointing a camera inadvertently at an overly bright light source. Several manufacturers—Sony, RCA, and Bosch, to name a few—have prototype cameras that include, attached to the back of the camera body, a small, lightweight VTR using current consumer product tape cassettes.

VIDEO CAMERA SETUP

In our discussion of video cameras so far, a number of controls and switches located on both the camera and its associated control unit, the ROU or CCU, were mentioned with little or no explanation of the controls' functions. We saved those explanations for this point in the text where we describe *setup.*

This setup procedure differs somewhat from camera model to camera model and to the degree that it must be repeated on the same camera. We start, to make the setup description complete, with a new camera, just out of its packing crate.

Hitachi FP-21 Camera

Here is the initial setup for the Hitachi FP-21 camera:

Unpack the camera, place it on a tripod, and connect its viewfinder and other cables properly.

Light the studio to its normal operating illumination if setup is to be performed indoors.

Turn on the camera power, and begin a 30-minute warmup of the heaters in the camera pickup tubes, with the power operate/standby (OPE/STDBY) switch in the STDBY position. This warmup protects the camera's pickup tubes and prolongs their service life.

After preheating, perform a color bar check. The color bar signal is available at the CAM OUT connector when the power operate/standby switch is set to OPE (operate) and the CAM/BAR switch is set to BAR.

Check the color bar vectors in the color bar signal with a vectorscope for precise chrominance; if necessary, adjust these parameters on the camera or its remote control unit.

Perform a video signal check. Obtain a color picture at the CAM OUT connector by switching the CAM/BAR switch to CAM and the filter disk to position 1 (3200 degrees Kelvin) or 2 (5600 degrees Kelvin), depending on whether the camera is indoors or outdoors. Check the video level and sync levels on the waveform monitor.

Check the white balance with the high-gain switch set to 0 db and the filter disk remaining at the color temperature of the lighting source. Focus on a pure white subject, such as a bedsheet, the back of a white shirt, or a white card, so that the subject fills the screen. Avoid having high-intensity light sources or their reflections shine on the white subject. Set the automatic black/automatic white switch to its upper position, and white balance is set in 0.5 second. The white-balance (W-BAL) indicator inside the viewfinder hood illuminates.

Check the black balance. Set the automatic black/automatic white switch to its lower position. The lens iris closes for 5 seconds, and black balance is obtained. After setting black balance, repeat the steps for white balance.

Both black balance and white balance are held in memory and need to be repeated only when the memory battery is changed (about every 2 years) or, in the case of white balancing, when the color temperature of the lighting has shifted. Color temperature shift occurs whenever the camera is moved from indoors to outdoors, or vice versa.

The pulse-canceling adjustment is performed by removing the camera's left panel, capping the camera lens with filter disk no. 4, setting the high-gain switch alternately in 0- and +18-db positions, and adjusting shading board controls RV 3, RV 4, and RV 5 until there is no change in the black balance or black level. When you adjust the pulse canceling, operate the master black level control as required. Upon completion of this adjustment, repeat the automatic black balance.

Now make the lens tracking adjustment. Return the filter disk to an uncapped position, and zoom the lens to TELEPHOTO. Adjust the lens focus for a sharp picture. Zoom the lens back to WIDE ANGLE, and check to see that focus is still sharp. If it is not, readjust, using the tracking control on the lens. If the lens has no tracking control or if the focal point differs among red, green, and blue images, then remove the left camera panel, loosen the R/G/B assembly lock screws at the front end of the camera head, and

adjust the tracking controls for optimum focus.

To make the registration adjustment, connect a picture monitor to the MON OUT connector. Focus the camera on a registration chart so that it fills the entire screen. The chart is used to check and adjust the camera for optical, mechanical, and electrical registration, so that the output signals are in correct time relationship. The chart uses a ruled grid pattern to aid in adjusting scan height, width, rotation, skew, centering, and linearity in each color channel.

Accurate registration requires that zone 1, within a circle having a diameter equal to 80 percent of the picture height, be less than 0.1 percent inaccurate; zone 2, within a circle having a diameter equal to the picture height, be less than 0.2 percent inaccurate; and zone 3, outside of zone 2, be less than 0.4 percent inaccurate. If the centering deviates in the center of the picture, adjust the red horizontal, blue horizontal, red vertical, and blue vertical centering controls. If the registration deviates because of picture corner distortion or picture tilt, adjustments are made on the deflection (DEF) card within the camera. The MON OUT VIDEO switches R/B and $G/-G$ are used to select $R - G$ and $B - G$, as appropriate, to check registration on the monitor.

Sony DXC 6000 Camera

These are the initial setup adjustments for the Sony DXC 6000 camera:

Check to see that the camera is properly connected, with the right cables to the device(s) (VTR, CCU, monitor) to which it will be feeding its video signal. On the camera, set the camera/VTR switch to ON/STBY; this supplies power to all parts of the camera. Set the gain switch to 0 and the output switch to CAM. Set the W/B mode select switch to AUTO/CCU and the CENT switch to REMOTE.

Set the automatic/manual switch to AUTO.

Turn the power at the power supply to ON and the camera power switch to ON. The viewfinder raster should appear in a few seconds. If it does not, turn up the BRT control on the viewfinder.

Remove the lens cap, point the camera at a subject, and turn the focus ring to focus the camera.

Perform the flange focal-length adjustment (back focus) which ensures that the subject is in focus at both the wide-angle position and the telephoto position during zooming. Set the automatic/manual switch to MANU, and set the iris ring to 16. Loosen the Ff (flange focal) ring lock screw. Zoom the lens to a telephoto shot. Turn the focus ring until a subject about 10 feet away is in focus. Zoom to a wide-angle shot, and turn the Ff ring until the same subject is in focus. Then repeat the telephoto and wide-angle zooms with the Ff ring adjustment until the subject is in perfect focus at both extremes, and tighten the Ff ring lock screw. Readjustment will not be necessary as long as the lens remains on the camera.

Iris adjustment is made with the automatic/manual switch in the MANU position. Focus the camera on a high-contrast scene or on a subject against a bright sky, and turn the iris ring. With the switch set to AUTO, the iris will be automatically adjusted to the brightness of the subject.

There are four settings on the filter wheel: setting 0 is blind (capped); setting 1 is 3200 degrees Kelvin, to be used with iodine lamps or outdoors at sunrise or sunset; setting 2 is 5600 degrees Kelvin plus ¼ neutral-density filter, for bright, sunny outdoor shooting; setting 3 is 5600 degrees Kelvin, for cloudy or rainy outdoor shooting.

The parameters for white and black balance should be adjusted for lifelike color and clear picture, under the lighting conditions that will occur while shooting and with the same filter that will be used. Focus the camera on a pure white subject, filling the screen. Set the W/B mode select switch to AUTO. Set the white/black switch to WHITE (the switch automatically returns to the center position when released), and wait momentarily until the W/B lamp in the viewfinder lights. Set the white/black switch to the BLACK position, and wait until the W/B lamp lights. Repeat the white-balance adjustment. If the lamp does not light, check that the filter wheel is turned to the right filter. Once white and black balances are adjusted, the adjusted values are memorized by the camera and held for about 3 days, whether or not there is power on the camera.

The video monitor adjustment is made by using the color bars generated in the camera. Set the camera output switch to BARS. Adjust the color and hue controls on the monitor while viewing the monitor screen. Set the output switch to CAM.

AESTHETICS OF CAMERA OPERATION

Now that we know how to set up the camera, let us forget that the camera is an electronic device and begin to look at it as an *artist's tool*. Let us look at how the camera is used to make programs or to gather news pictures.

First we examine how to put the camera in motion and learn the terms used for camera motion. If we swing the camera in an arc, or half circle, we say we are *panning*, from the word *panorama*. To pan smoothly, the cameraman first swings his body (with both feet planted firmly) in an arc in the opposite direction from the desired pan. Then, with the camera held on the shoulder or with the pan-tilt handle of a tripod-mounted camera held firmly, the operator's body swings back (uncoils) in the direction of the pan until the pan is complete and the body and feet point in the same direction. Try it.

If we aim the camera up or aim it down, we are *tilting*. With the camera on the shoulder, the tilt is done with the knees and shoulders, slowly and smoothly. If the camera is mounted, we use the pan-tilt handle, also slowly and smoothly.

If we move the camera from one position or place to another, forward or back, we are *dollying*. Closer to the subject is dollying in, farther away is dollying out. Moving the camera from side to side is called *trucking*. To the left is truck left, and to the right is truck right.

If we move closer to or farther away from the subject *without* moving the camera, we are zooming.

If the subject becomes sharper or fuzzier, we are changing focus. Hopefully if the focus becomes fuzzy, it is intentional and done artistically.

Points of Dramatic Departure

If the camera is at eye level with the subject, the subject and the viewer are equal in "stature." If the camera looks up at the subject, then the subject is dominant. If the camera looks down on a subject, then the viewer is dominant. The normal viewing level is at eye level with a subject in closeup, in the top third of the frame.

The picture that is seen in the viewfinder is complete, with horizontal and vertical edges, which permits proper centering. The viewer sees an "underscanned" picture because there is a *cropping* effect on the video picture, by the viewing picture tube mounting or frame, which requires that enough space or headroom be part of the picture to compensate for the masking of its edges. The camera operator should always allow at least 10 percent of the picture area, horizontally and vertically, as a "safe" area, an area within which the important picture information is included. When cropping a picture, as in a closeup, the camera operator should retain sufficient headroom at the top of the picture that the subject's head is not cut off on the viewing screen.

Instructions to the camera operator about the size of the shot are often found in the shooting script, and these instructions include the following (abbreviations in parentheses):

Long shot (ls)—a shot of the entire scene or set.

Three shot—three people.

Two shot—two people.

One shot—one person. As with two or three people, it can mean a full-length shot or a shot of head and shoulders.

Close-up (cu)—a tight shot in which the head or head and shoulders of the subject fill the screen.

Extreme close-up (ecu)—a very tight shot, often of one ear or one eye, that is done for dramatic effect.

CAMERA MOUNTING

The camera can be mounted on the cameraman's shoulder, which allows for all the flexibility of the human body in its motion. The disadvantage is that after a while the camera's weight will detract from the operator's ability to concentrate on the aesthetic quality of the pictures. And even though the camera is designed to be shoulder-mounted and has a soft, contoured shoulder pad, it is much more stable on a fixed mount. Recall that the longer focal lengths of a zoom lens magnify not only the picture but also any shaking or instability. As a rule, when the camera is hand-held, the camera operator does not zoom in too tight or zoom out too far, to avoid instability.

The camera can be mounted on a pan-tilt head, a device that can be swung from side to side and up and down, by using its pan-tilt handle, with all these directions mechanically *damped* to provide smooth motion. The pan-tilt head can be mounted on a three-legged device called a *tripod*, with or without wheels underneath; this is used primarily in the field. Or it can sit atop a studio camera pedestal (not to be confused with pedestal black) or a studio dolly where it can glide horizontally around on wheels and be cranked up and down a short distance. Often the studio dolly at network broadcast stations is

moved by a dolly operator, which leaves the cameraoperator free to frame and focus. (See Fig. 3–28.)

The camera can be mounted alongside the cameraman on a crane, which can move to heights of 10 feet or more above or move down to near floor level.

However the camera follows the action of the program, it does so either on the initiative of the cameraoperator or at the direction of the program's director. The cameraman maintains focus, frames the picture so that the subject is always the primary point of viewer interest, and moves or zooms to follow the action of the scene.

Before we discussed individual manufacturer's cameras, we mentioned that the camera could feed into a VTR directly or could be combined with other cameras, in fact with other sources of video, to be mixed into a program. Indeed, when we looked at those cameras, their system diagrams noted such combinations. Also, in most camera systems a microphone could be mounted to the camera, and the audio

Figure 3–28A Pan-tilt head. *(Provided courtesy of Quickset.)*

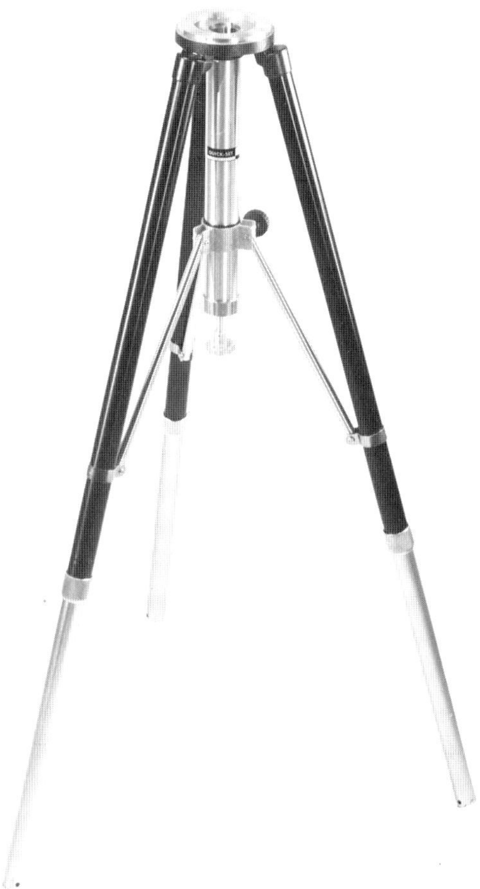

Figure 3–28B Tripod. *(Provided courtesy of Quick-set.)*

Figure 3–28C Trollie. *(Provided courtesy of Quick-set.)*

signal could be fed into the production system via the camera cable. The direct camera method of audio pickup is not recommended by this author. We discuss audio in detail in Chapter 5, including camera intercom as well, but we did not want to conclude our discussion of cameras without acknowledging the existence of the availability of audio from the camera.

CAMERA CABLE ABUSE

The signal—video or audio—gets from the camera to its destination via camera ca-

bles and cable connectors at the ends of those cables (Fig. 3–29). Camera cables and connectors require some discussion here because they receive a considerable amount of physical abuse and because no matter how one preaches to the operating crew, the abuse does not abate. Cables are stepped on, they are twisted, and they are run into by heavy wheeled pedestals and microphone booms. Connectors are dropped to concrete floors, are squeezed out of round by heavy feet, or suffer attempted forced mating when their pins and sockets will not coincide, resulting in bent pins. The danger is that cables and their connectors may fail to operate at just the instant when the irreplaceable pictures are being shot. An old rule, practiced by wise technicians throughout the industry, applies here: *Treat your equipment with respect.*

Figure 3–28D Pedestal. *(Provided courtesy of Quick-set.)*

Figure 3–28E Crane. *(Provided courtesy of Listec.)*

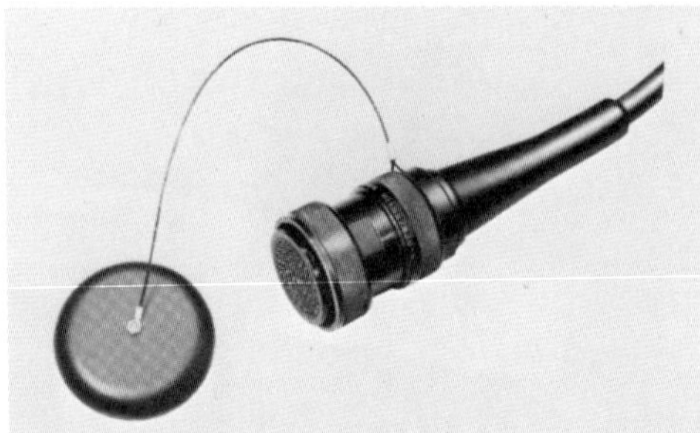

Figure 3–29 Camera connector. *(Provided courtesy of RCA.)*

We discuss cables and connectors in detail in a later chapter.

REVIEW QUESTIONS

1. What is a camera tube target? Of what materials is it made?
2. Describe scanning.
3. What is a frame? A field?
4. How many lines are there in the NTSC system?
5. What is the sweep frequency?
6. What is the picture aspect ratio?
7. Describe a raster.
8. Discuss the vertical blanking interval.
9. Describe composite video.
10. What is a color subcarrier? Why is it used?
11. Why is a waveform monitor used? A vectorscope?
12. What are color bars?
13. Describe genlock.
14. Explain luminance and chrominance.
15. What is camera optics?
16. What is measured in degrees Kelvin?
17. What is the purpose of camera setup?
18. What are a ls, a cu, and an ecu?
19. On what are video cameras mounted?

Chapter 4
The Production Switcher

PRODUCTION SWITCHER

When a camera signal is to be combined with other video sources, it is fed into a *production switcher* (Fig. 4–1). As with most major components of television production equipment, switchers range widely in their complexity and accordingly in price. Production switchers used in cable television generally fall into the $5000 price category, which is the lower range in price and complexity. Switchers may have few or many *inputs*, with generally two outputs, one preview output and one line or program output, which feeds the program to its destination.

The functions of switchers include the combining of camera and/or other incoming video inputs with the insertion into the picture of self-generated special effects (Fig. 4–2). A switcher may be described in terms of its number of inputs and outputs, such as a 7×2 and an 8×2, which indicates that it has 7 inputs with 2 outputs and 8 inputs with 2 outputs, respectively.

Combining video signals is performed electronically in the switcher by moving the video signal to and from *buses*, or switch points. Electronically, a *bus* is a line extension of a single point. A switching bus is a series of switch points, inputs, or intermediate points, only one of which can be switched to the bus output at any given time.

Thus, in our basic illustration of switcher operation (Fig. 4–2), four cameras operating in a studio are connected to a switcher in the control room. These cameras are inputs 1 through 4 on the switcher, with each of the cameras, or inputs, displayed on *one* of a set of horizontal pushbuttons, on *each* of three input buses, on the switcher. The pushbuttons are labeled "cam 1," "cam 2," "cam 3," and "cam 4," on the first four push-to-operate switches of each bus. Each pushbutton illuminates when it is activated (pressed), and its illumination goes out when it is deactivated, so that its status can be determined instantly. The buttons

Figure 4–1 Drawing—production switcher. *(Provided courtesy of JVC.)*

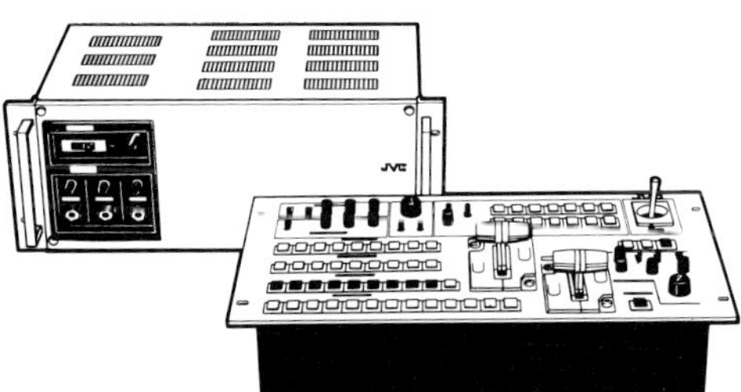

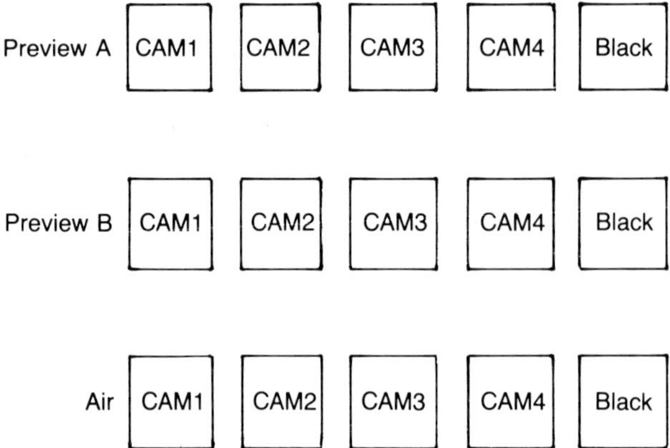

Figure 4–2 Basic switcher diagram—3 bus, switch-only.

on each bus are interlocked, so that only one input on a bus can be "on" at one time.

In addition, for the pushbutton switches to operate, the t-bar dissolver control must be at either extreme. We say more about this control later.

The buses, in turn, are labeled "preview A," "preview B," and "air" in Fig. 4–2, and each bus *output* (preview or air) has a monitor connected to it. So whichever video signal has been "punched up" on a bus can be seen on a monitor.

Cam 1, then, can be punched up on the "preview A" button. If its picture as seen on the preview monitor is wanted by the director, then a *technician* on the director's cue, by punching up cam 1 on the air bus puts its picture on the air (feeds it to a destination external to the control room) and also puts that picture on the air monitor of the control room. If cam 2 is punched up on preview B and is seen on the preview monitor, and if the director wants to go from cam 1 to cam 2, then the cam 2 air bus switch is punched up. It instantly switches the picture on the air bus from cam 1 to cam 2. The director

calls for, or cues, cam 2 in a two-step sequence: The first cue to the technician is, "Ready 2." The second cue is, "Take 2," or, better yet, "2 take."

Often, even though the technical director (TD) (switcher operator) has been given a standby for camera 2, at the last second the director may want camera 3. Waiting for the cue, the switcher operator is poised with a hand over the camera 2 pushbutton. And at the instant that he hears "take," the switch is made. If the cue is "3 take," there is enough hesitation for him to move his finger from button 2 to button 3, to give the director the camera he finally wanted. The term *take* is sometimes varied by the use of the term *cut*, instead, as in "ready 3, cut to 3."

That is the basic essence of *switching*. There is another way, however, of using the switcher in making the transition from one video signal to another, in this case, from cam 1 to cam 2, on the air.

Lap Dissolve

This second method is called the *dissolve*, or *lap dissolve*, where the cam 1

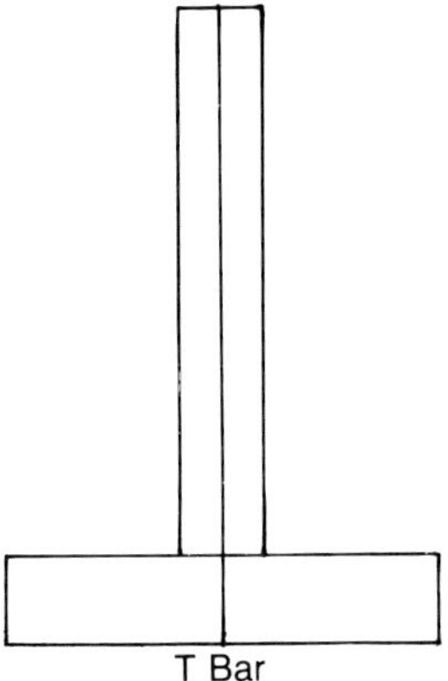

T Bar

Figure 4–3 The T bar dissolve control drawing. *(Provided courtesy of ISI.)*

picture dissolves or fades out as the cam 2 picture fades in.

To accomplish the dissolve, first a pushbutton near the fade control is activated on some switchers, which transfers the transition from switch to lap dissolve. When the T-bar handle is moved up, or down, from one extreme to the other, the picture on the air fades from cam 1 to cam 2. The fade can be fast or slow, and it can be stopped at any point for a superimposed picture of both camera signals. Typically, "supers" (superimpositions) are done on two differing camera angles of the same subject.

Close observation will show that the T-bar handle is *split;* it is actually two handles, one controlling each camera, working in opposing directions. So when one-half of the handle fades in cam 2, the other half fades out cam 1. The director cues this move with "ready to dissolve to 2" (or "ready lap 2"), followed by "dissolve to 2," or, as indicated earlier, "2 dissolve."

In addition to making the transition from one video signal to another, the production switcher has a built-in mix-effects generator with control(s) which allows the operator to select either none or one of a number of screen patterns. This permits *two or more* signals to be displayed on the screen at the same time. The screen may be "split," vertically, horizontally, or diagonally. One picture may be displayed in a box, circle, rectangle, or diamond, by a *wipe* within the first picture; and a joystick on the switcher adjusts the position of the internal picture precisely.

The tally system, indicating which camera is "on," is a part of the switcher function and occurs automatically when the transition from camera to camera is made.

The switcher, in addition to the buses previously described, may have a combination of perhaps three other buses which are used to display and mix special effects or to "key" (superimpose) titles or internally generated colors from a colorizer or matte colorizer generator to the picture background.

The one color that is essential from the color generator is black, as a fade or transition to program limbo. The director's cue is, "Go to black."

The special-effects combination can be switched to one of the preview or air buses. Bus outputs can be made to reenter the switcher for a combining effect. Further switcher refinements include preset combinations of mixing, switching, and wiping. The number and complexity of the special effects available are determined by the price of the switcher.

Another way to describe the operation of a switcher is to show it, in this case, as a 7-by-7 switch matrix (Fig. 4–4). In the diagram, each X represents a switch point or pushbutton, and each horizontal line of X's represents a bus. The first bus (line 1 across) is the cut, or take, bus, which "takes" any input signal or the output of the special-effects or lap dissolve sections of the switcher (button 8 or 9).

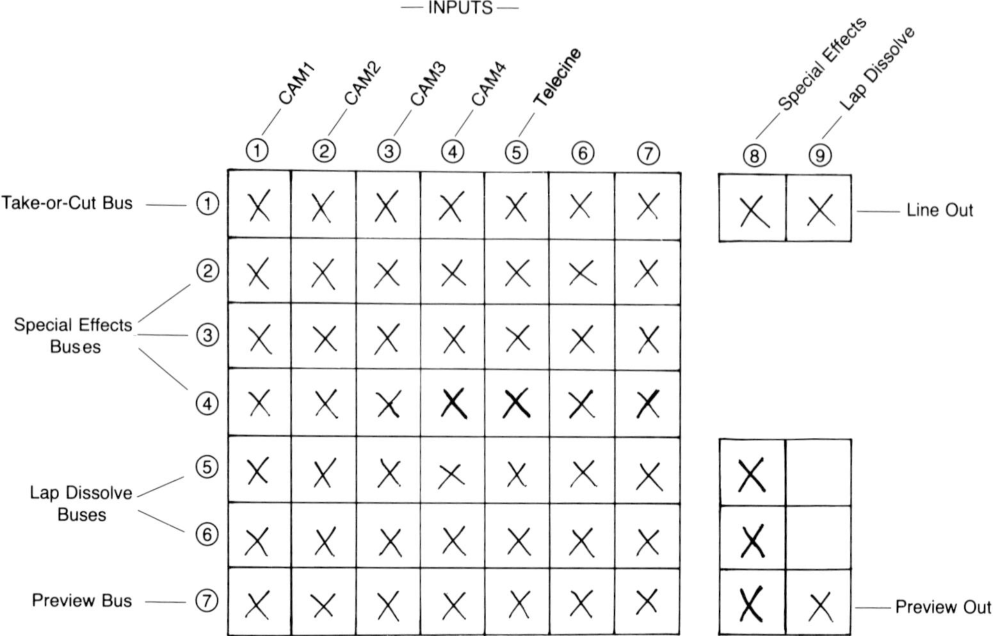

Figure 4–4 7X2 Switcher matrix diagram.

The next three buses (lines 2, 3, and 4 across) are used to create special effects or to key titles, color, or other effects. The special effects or key effects are switched to the line output when switch point 8 is pressed on the top row, or take bus. The same special effects or key effects can be switched instead to the lap dissolve mixer by pressing button 8 on whichever lap dissolve bus (horizontal line 5 or 6) is in use. Then the lap dissolve mixer can be switched to line output by pressing button 9 on the top row. And the effects buses can be switched to the preview bus (line 7) for viewing while the desired effect is put together.

The Transition Rationale

Every technical director develops individual operating patterns and must learn to operate each new switcher differently. What *always* remains the same in switcher operation is the motivation behind each transition operation. Whether one uses a take or a dissolve depends on the mood or effect one wishes to create. Time lapse is better described with a dissolve than with a cut. The motivation for a special effect—such as a *wipe*—which replaces one picture with another, in a horizontal, vertical, or diagonal direction, or an *iris* where one picture is replaced with another from the center out or from the edges in toward the center—should always begin with a valid reason, not simply because these special effects, or a variety of others are "there," and available from the effects generator of the switcher.

Transitions *always* occur on a cue to the audience. They can be thought of as existing in one of two frames of reference, either in "time" or in "place," and they can occur on an audio cue or a video cue or both. Here are some examples, but remember that it is awkward, at best, to try to describe video motion in words:

- *With an audio lead cue.* The audio starts in one scene's end and continues in the next scene.

- *On a verbal cue.* This is done when someone explains where she is going and the cut is to the next scene which finds her there. *Example:* "I'm going shopping," and the next scene is a shopping center. Or reference is made to another scene, and the switcher cuts to the referenced scene.

- *On an object cue.* The telephone rings, someone picks it up, and the scene cuts to the person at the other end. Then the camera picks up the latter scene. A variation is the phone ring which triggers a split-screen cue depicting both ends of the conversation.

- *With a music cue.* The pictures are placed gauzelike, over music, and the music transcends the pictures and brings them into another place or another context.

- *With an establishing shot change.* The outside of a structure is shown; on cue, the picture switches to action within the structure.

- *On a dissolve.* A dissolve from one scene to another demonstrates a lapse of time, either forward or backward in time, as in a flashback. In a flashback, the scene is often started with a modulated key, which causes the picture to waver or oscillate vertically.

- *On a surprise change shot.* The camera will cut from an outside shot of someone throwing an object into a window to a scene inside the thoughts of the thrower.

- *On a theme transition.* A driver gets into a car and drives away left. The scene cuts or dissoves to the car driving to the right.

- *The "follow in."* A wide shot of a strange scene focuses on a stranger entering the scene and brings the camera, following the stranger, to a table where he interacts with known characters in the scene.

- *The focus in/focus out.* By using very short depth of field, the camera lens focuses sharply on one of two interacting players, with the other player out of focus. As the second player speaks, the focus shifts and reverses, sharp on the second player and soft on the first.

- *Intercuts.* Portions of two separate taped interviews are intertwined so as to suggest a debate when, in fact, none existed.

- *On an action cue—backstroke to backhand.* The backstroke of a swimmer cuts to the backhand of a tennis player.

- *On narration.* "What you are about to see . . ." helps in the video transition of scenes. A classic example is any one of the old Jack Webb police dramas.

- *Video lead-in.* The video picture briefly leads the audio, which is still in the previous scene.

- *Fast cuts in very rapid succession.* These highlight an event or an activity without showing every aspect.

- *Action reversal.* A verbal cue such as "help me up," with the camera tightly depicting one person helping another up from the ground, reverses on a cut to show the person being helped. Or the second person may be shown being helped in another location and in a different situation.

Let us now consider a couple of brand-name switchers that are used in television.

Crosspoint Latch 6114 Switcher

The Crosspoint Latch 6114 switcher has eight video inputs to three buses (preview A, preview B/key, and program). (See Fig.

4–5). There are two program channel outputs and two preview channel outputs. The preview channel, with two buses, is used for advance setup of effects, including preset wipes and pattern modulation. Input no. 8 is connected to an internal colorizer, a generator which provides color or black for background keying. The other seven inputs are from external video sources. All switching is done during the vertical interval except for effects pattern selections.

The 6114 has a 12-pattern mix-effects generator which may be switched into either the preview or the program output. A joystick positioner, when activated, allows positioning of five effects. Effects can be further enhanced by built-in soft-wipe and spotlight controls and a pattern modulator with frequency and amplitude adjustment. Keying and color matting may be done by an external video input or by selecting one of the inputs on the preview B bus, which doubles as a key bus. Wipes can be either normal or normal/reverse, by selecting the appropriate operational mode.

A preset control allows for the predetermined movement from mix-to-preset-wipe and wipe-to-preset-mix operations. An AUTO MIX/WIPE button automatically mixes or wipes to a new input.

The model 6114 has a digital clock/timer with hold and reset controls. There is also a camera tally system.

Operating Controls. At the upper left corner of the switcher panel is found the *pattern selector.* To its right is the SOFT switch, which in the ON position provides a soft-edged wipe of all the effects. The term *soft* refers to the edge(s) of the effect, which can be either hard or diffused. Below the SOFT switch is the spotlight switch (SPOT) which turns on the spotlight effect when button no. 9, the direct program button on the program bus, is pressed and a wipe mode is selected. The spotlight effect enhances the luminance, or brightens a selected portion of the picture.

The *blink key* switch to the right of the spotlight switch, in the ON position, will flash the selected pattern or key signal on and off. The modulator switch (MOD) to its right turns on the pattern modulator. Above it, the *level* knob controls the intensity of the modulated pattern, and the frequency knob (FREQ) controls the modulation speed.

The *event timer* is a digital clock controlled by the *timer clock* switch; it can run up to 24 hours. It has a three-position toggle switch: *start, stop and reset.* It can be set to the START/STOP position and is spring-loaded in the RESET position, to return to RUN in the center position.

To the right of the timer are the *color/black* controls. The two-position toggle switch selects either color or black background key, generated internally. The lu-

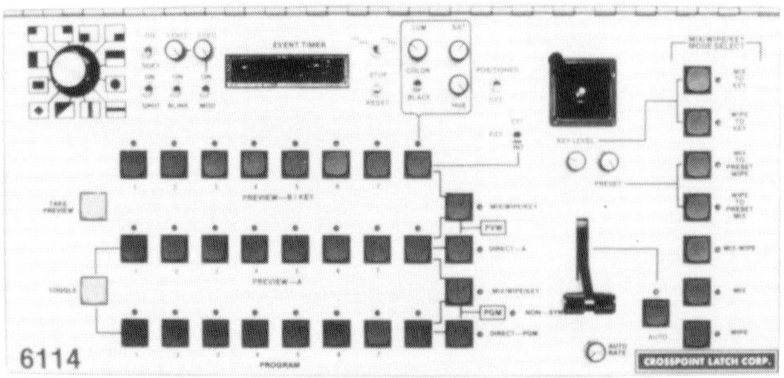

Figure 4–5 Crosspoint Latch 6114 video production switcher. *(Provided courtesy of Crosspoint latch.)*

minance (LUM), saturation (SAT), and hue (HUE) control knobs determine the color output of the background generator.

The *positioner* toggle switch activates and deactivates the *joystick*, which in turn controls the position of the selected effects.

Seven vertical pushbuttons on the right side of the control panel allow selection of any *one* mode of *mix, wipe* or *key*. An indicator light to the right of the pushbutton is illuminated when selected. The functions of this switch matrix, top to bottom, are as follows:

MIX to KEY, with level set by the KEY level control knob

WIPE to KEY, level set by KEY knob

MIX to PRESET WIPE, level set with preset control

WIPE to PRESET MIX, level set with preset control

MIX/WIPE, mix/wipe is selected

MIX, mix is selected

WIPE, wipe is selected

AUTO, the pushbutton to the left of and between the MIX and WIPE selectors, operates in only the mix, wipe, or mix-wipe modes and will not operate in any of the preset or key modes. Pressing the pushbutton automatically dissolves or wipes. The rate of the dissolve or wipe is controlled by the AUTO RATE control below.

The T-bar fader handle controls the mix effects manually.

Key is selected either internally on the preview B bus or externally by the KEY-EXT-INT switch.

In the *preview B key bus*, the first seven pushbuttons on the uppermost horizontal row select any one of seven inputs. The eighth button selects either color or black, determined by the COLOR/BLACK switch.

The *preview A bus* and *program bus* are the same as the preview B bus except that there is no internal key.

Adjacent and to the right of the input buses, there are four vertical pushbutton switches: MIX/WIPE/KEY (PVW), which permits performance of mix or wipe between preview A and preview B, as determined by the mode selector; DIRECT A, where preview A is a direct output; MIX/WIPE/KEY (PGM), which is the same as the MIX/WIPE/KEY (PVW) except that it functions between preview A and program buses; and DIRECT-PGM which puts the program bus on direct output.

The nonsync warning light indicates that a nonsynchronous input has been selected on a preview bus.

Adjacent and to the left of the input buses are two pushbuttons. The TAKE PREVIEW button switches selection from preview to program. For example, if the selection on the program bus is on DIRECT and input 1, preview A is on input 2, preview B is on input 3, the PVW mode is MIX/WIPE/KEY, then upon depression of the TAKE PREVIEW button the program bus will switch to input 2, preview A will switch to input 3, and the program mode will go to MIX/WIPE/KEY. An effect may be previewed between the preview A and preview B buses and then can be directly transferred to the program output by pressing the TAKE PREVIEW button. This button also serves another function. When an input is on the program bus and two other inputs on the preview A and B buses are selected, these preselected inputs can be transferred sequentially to the program bus by pressing the TAKE PREVIEW button when the switch is desired; the second switch, the TOGGLE button, causes switching, or toggling, both ways, between the program and preview A buses. The toggling action either occurs automatically at the end of a mix or effects transition or can be manually activated by pressing the TOGGLE button.

Mix Effects. The mix-effects circuitry of the 6114 can be switched either into the preview or program signal paths or into both at the same time. These signal paths share the pattern generator and the T-bar fader handle, but otherwise are quite independent. For example, it is possible to have a mix-effects setup between camera 1 and camera 2 on the program output and at the same time have a similar setup between camera 2 and camera 3 on the preview output. Then the preview output can be transferred to the program output by pressing the TAKE PREVIEW button.

The functions of mix-wipe, mix, wipe, wipe to preset mix, mix to preset wipe, wipe to key, and mix to key can be performed by the mix-effects circuitry. These operations may be controlled manually by using the T-bar or automatically by using the AUTO button. A light above the AUTO button, when illuminated, indicates that AUTO is available. The AUTO button is disabled when the T-bar is at other than one of its extreme settings, and when it is in any of the preset or key modes. This light serves as a tally indicator, by going out when the AUTO button is pressed.

Wipes. All wipes are normal in direction. That is, a circle will always wipe open outward, from the inside of the circle. If a normal-reverse wipe is wanted—that is, the circle opening outward when the fader handle is moved one way and the circle closing inward when the handle is moved the other way—it can be achieved by setting the PRESET control knob to the fully clockwise position and selecting wipe to preset mix. In this mode

the buses will not toggle and the normal-reverse wipe will be achieved.

Keying. Keying may be performed with either external or internal key signal sources. When an external source is used, the video level may be adjusted on a preview channel and maintained during switching to a program channel.

When an internal key signal is used, the following procedure should be employed:

1. Set the KEY switch to internal (INT), and select the key source to the preview B bus. Select the background on the preview A bus, and set the program bus to DIRECT.
2. Depress the PVW MIX/WIPE/KEY button and the wipe-to-key mode button; then move the fader handle to its opposite limit.
3. Looking at the preview monitor, adjust the KEY LEVEL control for self-key, the correct degree of keying.
4. Select the desired background input on the program bus, and select the desired signal input to be inserted on the preview A bus, leaving the keying source on the preview B bus.
5. Depress the PGM MIX/WIPE/KEY button. Move the fader handle slowly to its opposite limit. The key will be wiped on the program monitor.
6. The key signal may now be wiped OUT by returning the T-bar handle to its other limit. Or it may be dissolved by first pressing the MIX-TO-KEY button and then moving the handle to its other limit.

When a straight mix or mix-wipe is previewed on the preview monitor, the display automatically reverts to the original scene when the fader arm reaches its limit, or when the AUTO effect is complete. On program, it is unnecessary to return the handle to its original position before doing a take.

Spotlight.   Spotlighting is available only on the program bus, which should be in DIRECT. Any input may be selected on the preview A bus while spotlight is in effect. A circle pattern is usually selected for spotlight, but any pattern may be used.

To use spotlight, select the circle pattern and press the PGM DIRECT button. Then place the SPOT switch in the OFF position, select WIPE on the mode selector, and place the SOFT switch in the OFF position.

Select *different* inputs on preview A and preview B buses, and press the PVW MIX/WIPE/KEY button. Adjust the fader handle and the joystick control to get the right size and position for the circle pattern on the preview monitor.

Set the SPOT switch to ON. Now a brightened area corresponding to the size and the position of the circle pattern on the preview monitor will appear on the program monitor.

Set the SPOT switch to OFF. If it is left in the ON position, a flash will appear on the program monitor each time a mix or wipe is completed.

ISI 902 Switcher

The Industrial Sciences, Inc. (ISI) switcher model 902 has ten inputs with tally (Fig. 4–6). Input no. 1 is normalled to the black burst generator/colorizer. There are two mix-effects systems with a shared pattern generator, which permits hundreds of standard and special wipes to be created from the seven basic patterns. The vertical wedge pattern, for example, may be rotated to form a vertical wipe from center.

The functions in each mix-effects system include mix, wipe, key, mix key, wipe key, and mix-wipe key. The key system in each mix-effects system provides matte key, key invert, and a three-input selector that chooses among self-key, external key, and chroma key inputs.

There are four output buses plus separate program and preview switching buses. These allow preview and take of mix-effects (M/E) bus 1, mix-effects bus 2, or the B direct bus.

Softness of border, or border width of the variable-wipe edge, is adjustable with

Figure 4–6 The ISI 902 switcher. *(Provided courtesy of ISI.)*

the edge control, while border color is also variable with the colorizer control. A sine wave modulator can be used on both M/E 1 and M/E 2 to create effects which vary with frequency and gain controls.

M/E 1 offers preset wipes and electronic spotlight with adjustable positioning. M/E 1 can be reentered into M/E 2, enabling mix or wipe behind keys and colorized mattes wiping over chroma keys. (See Fig. 4–7).

American Data 2103 Production Switcher

The American Data 2103 vertical-interval color production switcher is self-contained except for an external power supply. Its input selection pushbuttons are momentary-contact, illuminated, legendable lens cap switches. There are ten inputs, seven of which accept composite synchronous signal, one black/reentry input, and two downstream (see Fig. 4–8) program inputs for nonsynchronous composite, or for noncomposite signals.

Program output switching allows the operator to select the MIX/EFFECTS output or either of the two external nonsyncronous inputs. A rotary-switch preview selector can be used to preview the mix output, the effects output, or either of the two external inputs.

A black burst/color background generator is included as part of the switcher to provide fades or wipes to any color or to black. When it is used together with the matte keyer, the background generator will provide color insert keying.

REVIEW QUESTIONS

1. What is the purpose of combining video signals?

2. What are the three functions performed by a video switcher?

Figure 4–7 ISI switcher model 902 functional diagram. *(Provided courtesy of ISI.)*

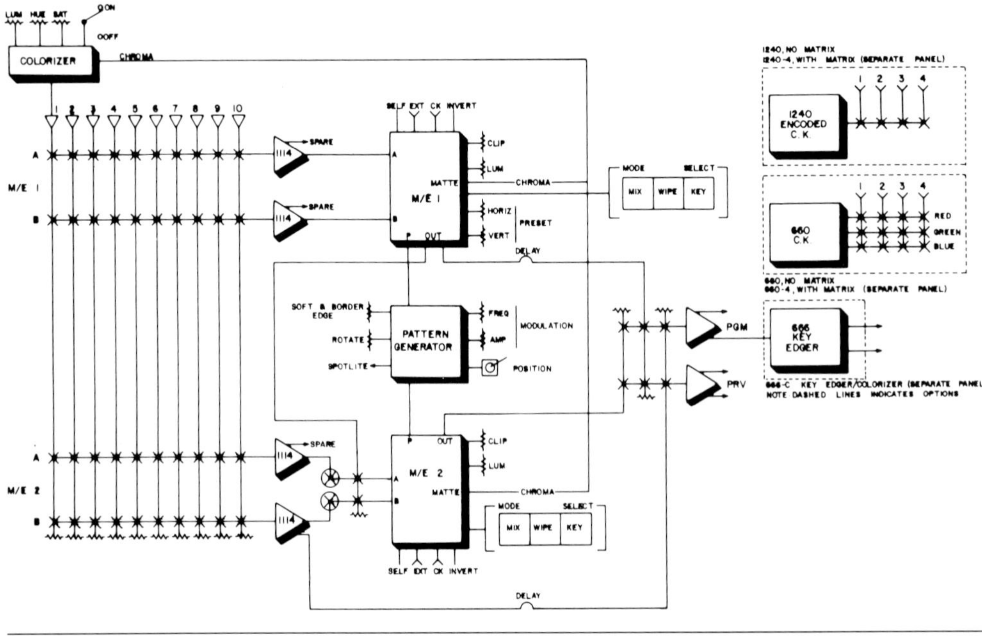

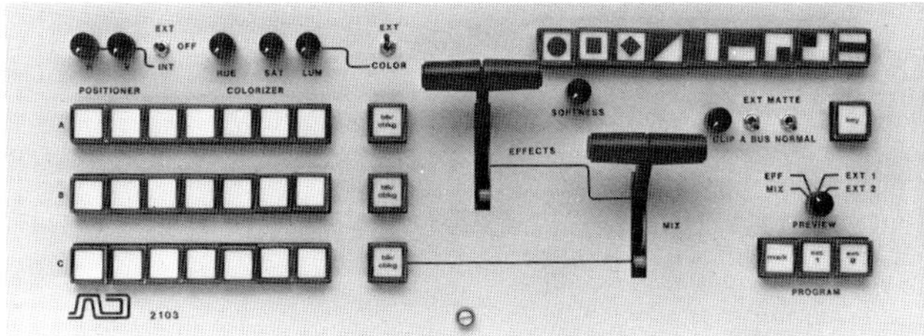

Figure 4–8 American Data 2103 switcher. *(Provided courtesy of American data.)*

3. What is a bus?
4. Why is the cue "2 take" better than "take 2"?
5. What is a dissolve? A super?
6. What is *the* essential color needed in video switching?
7. Why do we use special effects? When do we *not* use them?
8. Describe keying, a normal wipe, and a normal-reverse wipe.
9. What does SPOTLIGHT do?
10. What does a SOFT switch do?

Chapter 5
Television Audio

Audio, or sound, on television programming has always taken a secondary position to the picture. But clearly the two should be of equal importance. However, since the advent of the portable cassette recorder, with its built-in microphone, everyone who owns a cassette recorder feels that he has mastered audio by turning on the recorder. We show you here why it just is not so.

Most of the material in this chapter is taken directly from the author's earlier text, *Audio Control Handbook*, 5th ed. (Hastings House, Publishers, New York, 1983). Here, however, we limit the discussion to the audio equipment in use in studios and on electronic field production and news gathering excursions at access cable installations.

We talk here about microphones, or mikes (sometimes abbreviated and spelled *mics*), and mike systems as well as the mixers that blend the sound of two or more mikes and inject that sound blend into the TV program. We also discuss an audio system that links the cameraman in a studio with the program director in the control room, known as camera intercom. Again, as for those readers who want to delve deeper into lenses, much more about audio can be found in the text just mentioned.

DEFINITION OF SOUND

Sound, upon entering a microphone, is transformed from mechanical energy to electric energy. A device that can change energy from one form to another is called a *transducer*. Similarly, a loudspeaker, or earphones, which changes electric to mechanical energy is also a transducer.

Occasionally, in this chapter, we may, for convenience, refer to sound as traveling through electronic components, although we are well aware of the variance from scientific fact.

As a performer speaks, sound waves are created and radiate in all directions from the sound source. The waves impinge on many objects; some absorb, others reflect, and still others vibrate sympathetically with the sound waves.

A sound wave has a pressure component, or strength of the wave; a velocity component, or speed of the wave; and a frequency component, or cyclical/tonal characteristic, which is expressed in hertz (Hz), or cycles per second. Audible sound is heard by the human ear within the audio frequency spectrum of 20 to 20,000 Hz. [Note: 20,000 Hz = 20 kilohertz (kHz).]

With a microphone in close proximity, a portion of the sound wave strikes the transducing element, causing it to vibrate sympathetically and transform the sound energy to electric energy.

MICROPHONES

Microphones are described in terms of their directionality, with descriptive terms such as *unidirectional* (one-directional), *bidirectional* (two-directional), *omnidirectional* (all-directional) and with polar pattern graphs; and in terms of their frequency response, or place in the audio spectrum, with response curves which indicate their frequency range. There are also the "peculiarity factors" of each microphone type, which further narrow the mike's applicability to specific sound pickup applications.

Polar Patterns

A *polar pattern* is a concentric circle graph which is marked off at 0° (the axis), 90°, and 180°. The microphone pickup pattern is an overlay on the concentric circles that describes the mike's directionality. There are three distinct patterns, with variations existing in all three:

1. *Omnidirectional pattern.* Microphones with this pattern configuration accept sound with virtually equal facility from any direction. However, in some omnidirectional mikes, the pure circular pattern becomes somewhat egg-shaped as sound frequency rises toward the very high end of the sound spectrum. (See Fig. 5–1.)

2. *Bidirectional (figure-eight) pattern.* Mikes with this configuration accept sound with maximum facility at the 0° and 180° points, with minimum facility at either of the 90° points and with varying degrees of facility at in-

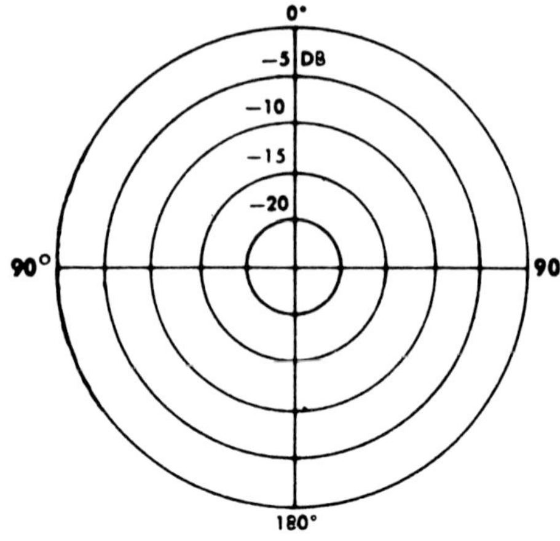

Figure 5–1 Omni-directional polar pattern.

between points on the graph. This pattern varies as the sizes of the lobes of the figure eight change in relationship to each other. That is, if either the upper or lower lobe is made larger or smaller, then the mike acceptance pattern changes. (See Fig. 5–2.)

3. *Unidirectional (cardioid) pattern.* Mikes with this polar pattern accept sound with maximum facility at the 0° point, with minimum facility at the 180° point, and with varying degrees of facility at in-between points.

Figure 5–2 Bi-directional polar pattern.

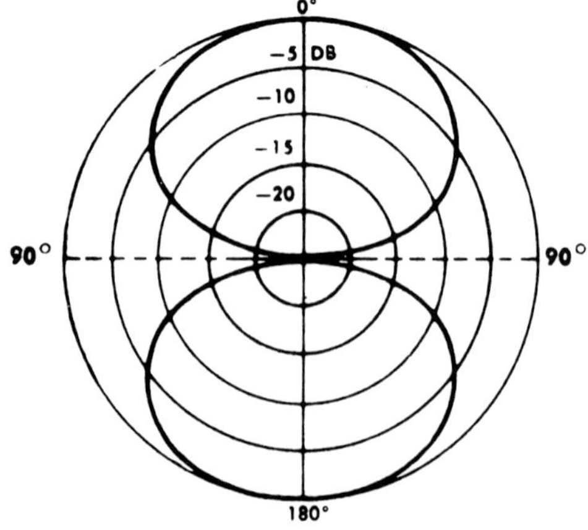

This pattern varies as the size and shape of the lobes projecting below the 90° points on the graph are changed. (See Fig. 5–3.)

In both the bidirectional and unidirectional microphones, the lobe shapes are design characteristics of particular manufacturers. In a least two older mikes, the lobe shapes can be varied by a recessed switch in the body of the mike.

The term *cardioid*, applied to the unidirectional mike, refers to its heart-shaped pattern.

Frequency Response Curves

As is true with polar patterns, the mike's manufacturer supplies a response curve for each mike as a part of its specification sheet. The curve is plotted on a graph of a typical x axis versus y axis, with decibels of gain on the vertical axis and frequency in hertz on the horizontal axis. In reading the graph, we look for the "flat" portion of the response curve, or for where the response is substantially equal, within a decibel or two, over the greatest frequency range. We also look at both ends of the curve to ascertain the degree of "peaking" or "roll-off" at each end. In other words, we look for the working portions of the curve.

Usually neophytes ask, "Why are all good mikes not designed with a flat curve throughout the audio spectrum (20 Hz to 20 kHz)?" The answer is twofold. First, this design would produce a very costly instrument—if it were possible. Second, the manner in which mikes are used requires that some have a particular roll-off or that others have a particular peak at specific frequencies, as a compensating factor. For instance, *lavaliere*, or *tie-tack*, *mikes*, which are worn on the performer's chest, below and beneath the chin, are designed with a particular high-frequency peak and a low-frequency roll-off to compensate for that particular mike placement.

Both the human ear and microphones respond to changes in the pressure component of a sound wave. The *sensitivity* of a mike, in decibels on its response curve, is the alternating voltage in millivolts at the mike output which results when a sound wave with a pressure of 1 microbar impinges on the transducer element. A sound source at normal listening level at a distance of about 2 feet is equal to about 1 microbar. Sensitivity is measured with the sound wave hitting the transducer on axis, or at 0°, unless it is stated otherwise on the response curve.

To compare the frequency response of on-axis sound impingement with that of sound arriving at the transducer at other angles (that is, to indicate the directional sensitivity of a mike), often other response curves are shown on the same graph, with the number of degrees off axis stated next to the curve. The difference between response at 0° and 180°, or be-

Figure 5–3 The uni-directional polar pattern.

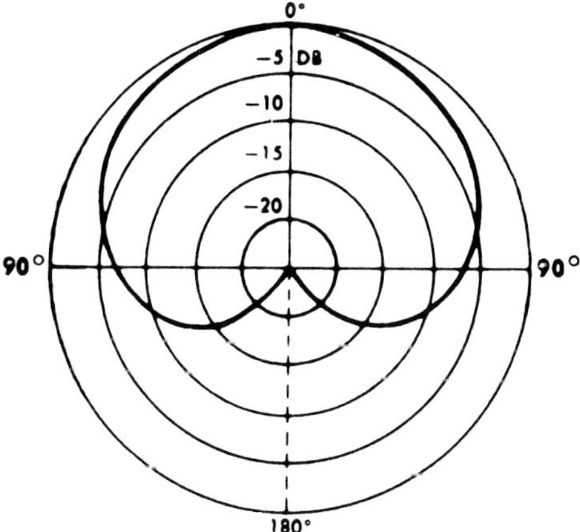

tween the front and the back of the mike, is termed the *front-to-back ratio.*

Reading a mike response curve, then, amounts to comparing measured ratios of sensitivity at differing frequencies along the audio spectrum, expressed in decibels. (See Fig. 5–4.)

"Peculiarity" Factors of Mike Types

These factors are described in terms of the idiosyncracies of particular microphone types.

The *dynamic mike,* because of its use of a plastic or metallic diaphragm transducing element, is the most rugged in physical construction. It is subject to the distortive effect of wind noise when it is used out of doors. But that problem can be measurably lessened by placing a foam plastic windscreen over the head of the mike. Dynamic mikes tend to discriminate in favor of high-frequency sound sources. This factor can be used to advantage when miking a speaker with a particularly deep bass voice, to eliminate some of the bass. Because of this sensitivity to the high end of the spectrum, dynamic mikes tend to accentuate the sibilance inherent in some voices. Sibi-

lance is the hissing sound made in the pronunciation of the letter S.

Dynamic *condenser mikes,* which use a capacitive transducing element called an *electret,* are much "hotter" mikes, have more output gain, than their plain dynamic counterparts. A special high-polymer film is charged with electrostatic energy under high-temperature conditions. The polymer film retains the charge almost permanently, even after its temperature has returned to normal. The charged film is called *electret,* and it is stretched across a ring to form a diaphragm which vibrates with sound. As it vibrates, the gap distance between the electret diaphragm and the mike's back plate changes, causing a capacitive change, which induces a small variation in the electric potential of the electret film. This variation in potential is then amplified by a built-in preamplifier and appears as the mike output signal.

The *velocity mike* in earlier versions was an extremely fragile instrument, but newer models are quite acceptable for indoor studio use. Velocity mikes are used for performers who have a tendency to "pop." *Popping* can be described as an overstressing of the explosive consonants B, P, and T. The explosive quality of these

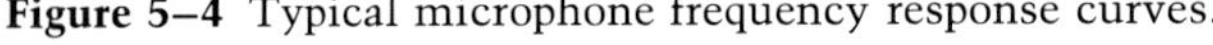

Figure 5–4 Typical microphone frequency response curves.

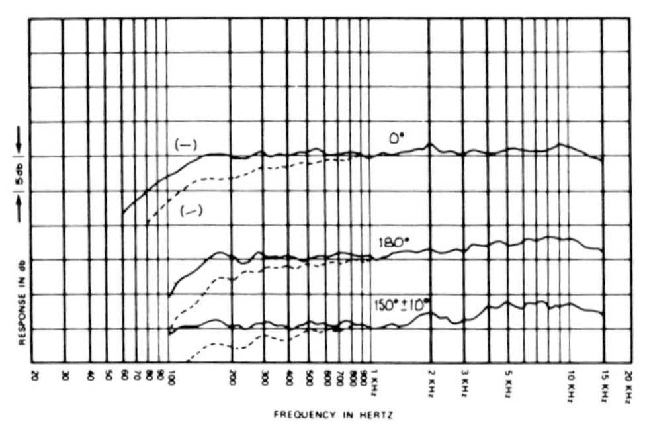

Frequency Response

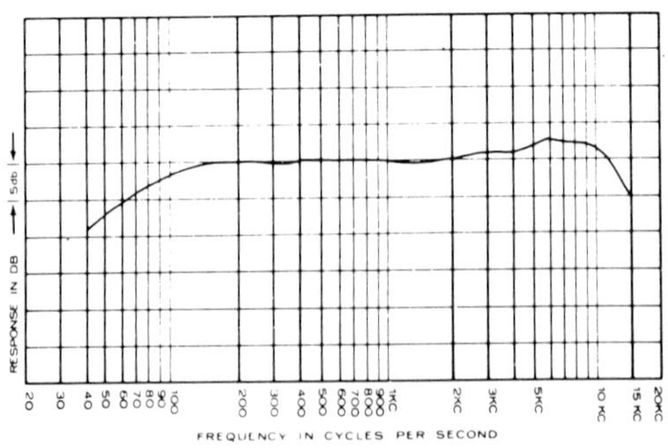

Response Curve

letters causes a sharp momentary rise in the pressure component of the sound wave. The velocity mike, often called the *ribbon mike* (because of the construction of its transducer), is actuated by the velocity component of the sound wave and so is less apt to be affected by popping.

Ribbon mikes are most sensitive in the lower range of the audio spectrum. Similar to the single-D cardioid mikes which we discuss next, ribbon mikes may be used to advantage to pick up voices that are too high in pitch. Furthermore, the closer that a performer works to a ribbon mike, the lower the pitch will appear to be. This effect is termed the *proximity effect.*

The *cardioid mike* appears in two distinct types, utilizing either a dynamic transducer or a condenser dynamic transducer. A cardioid unidirectional mike works as it does because its case is not sealed. In fact, the sound pressure is literally "invited" to contact the mike's diaphragm from its back as well as from the front, through a port in the back of the mike. The sound pressure from the rear, delayed by the distance it has to travel to the rear port and back internally to the diaphragm, neutralizes the diaphragm motion somewhat. This type of cardioid mike is called a *single-D*, for the single distance from the rear port to the diaphragm. The single-D has a frequency response which varies greatly with the distance between the source and the mike. At 0.25 inch from the sound source, the bass response is 15 db higher than the response if the mike is 2 or more feet away. This proximity effect is used extensively when it is necessary to add bass to a mike pickup and is avoided in heavy existent bass situations.

The Electro-Voice company, a major U.S. microphone manufacturer, has developed *variable-D* mikes with multiple ports, which reduce the bass-boosting proximity effect but maintain the unidirectional pattern.

Well, then, should we always opt for a cardioid mike rather than an omnidirectional mike? Here are the tradeoffs: Within the same price range, an omnidirectional mike has a smoother frequency response than a cardioid mike, is significantly less susceptible to breath noise and mechanical shock, and is often more rugged. But the unidirectional mike increases the possible working distance between source and mike to a theoretical 1.7 to 1. Clearly, the use to which a mike will be put determines which type should be chosen. Cardioid mikes are generally more expensive mikes, but that alone does not make them better.

If the use requirement calls for an even narrower cone of sound acceptance than can be achieved with the usual cardioid mike, then a cardioid variation called the *line mike,* or *shotgun,* may be used. Line mikes such as the Electro-Voice Cardiline have polar patterns that are even narrower than regular cardioids, and they are used for pickups at distances beyond normal mike ability, as on boom mounts in television studios or even while being hand-held on "fish poles" by electronic news gatherers. The polar pattern variations of the supercardioid variable-D and the Cardiline mike are shown in Fig. 5–5.

The directionality of a cardioid mike

Figure 5–5 Super-cardioid *(left)* and cardiline *(right)* polar patterns.

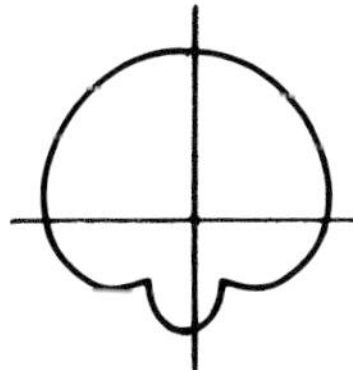

can be demonstrated by moving a sound source of constant level around the mike at a constant distance from the mike. Motion of the sound source in any direction away from the mike's specified angle of acceptance lessens the acceptance of sound. With the axis at 0°, a *cone of acceptance* is described, with the apex of the cone at the transducing element and the base of the cone formed at the moving sound source. Within the cone area (shown by the dashed lines of Fig. 5–6), if the distance between the sound source and the mike is increased—the sound source is moved farther from the mike—then the base of the cone becomes larger, while the angle of acceptance stays the same. But the amount of sound reaching the mike becomes less as its decrease follows the inverse-square law.

Now we examine some mikes used in television operation.

Electro-Voice 635A Microphone

The 635A is one of the most popular and most rugged mikes used in both radio and television (Fig. 5–7). It is a straight omnidirectional mike with an output of −55 db and a frequency response of 80 to 13 kHz. No polar pattern is shown because it exactly matches that of Fig. 5–1. It is used in hand-held situations where the sound source is close to the mike.

Figure 5–6 Cardioid mike cone of acceptance.

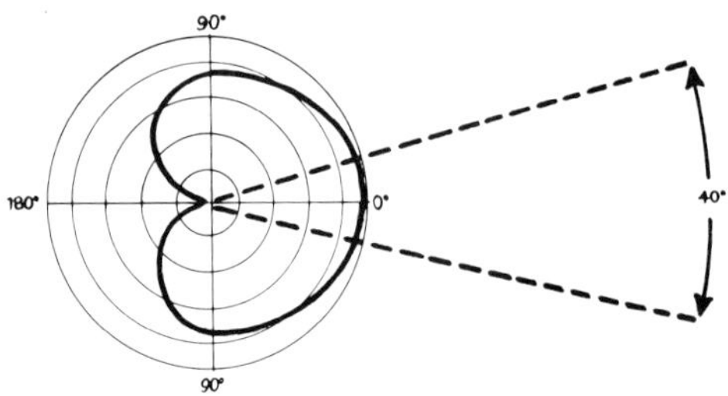

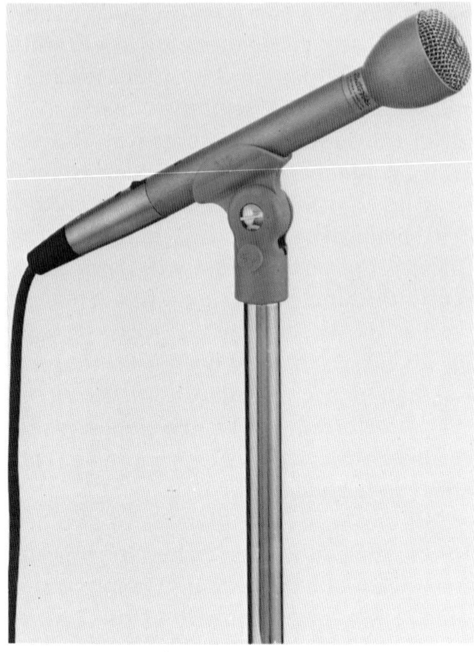

Figure 5–7A Photograph of E-V 635A microphone. *(Provided courtesy of Electro-Voice.)*

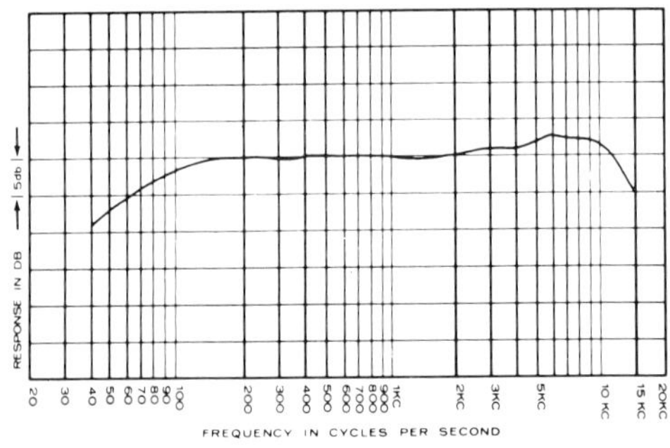

Figure 5–7B Frequency response drawing for E-V 635A microphone. *(Provided courtesy of Electro-Voice.)*

Electro-Voice RE-16 Microphone

The RE-16 is a dynamic variable-D supercardioid mike (Fig. 5–8). It is second only to the 635A in its popularity in the industry as a general-purpose mike. It has good sound rejection from the rear. Its fre-

Figure 5–8A Photograph of E-V RE-16 microphone. *(Provided courtesy of Electro-Voice.)*

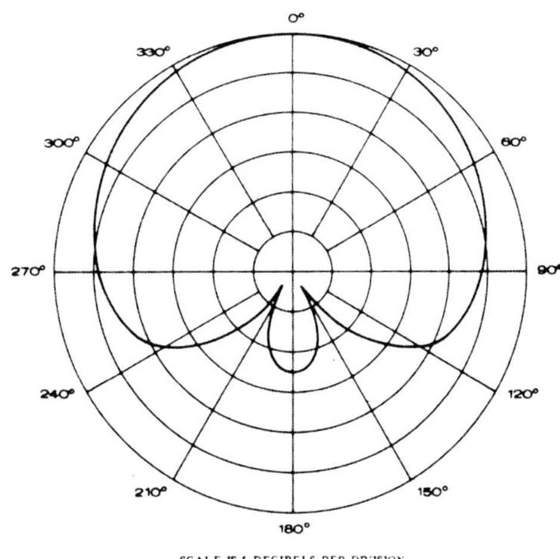

Figure 5–8B Polor pattern of E-V RE-16. *(Provided courtesy of Electro-Voice.)*

quency response of 80 Hz to 15 kHz is unusually independent of the angular location of the sound source. Switchable roll-off below 150 Hz keeps low-frequency noise to a minimum. It has an output of −56 db and a built-in blast filter.

Sony ECM-50PS Microphone

The ECM-50PS was the first small tie-tack mike to replace the larger lavaliere mikes which hang around the neck. It is an omnidirectional electret dynamic mike. The suffix PS indicates that its internal preamplifier may be powered in two ways: either by a built-in (to a belt-clip battery box) battery supply or from a simplex (phantom-powered) system which sends operating power through the mike signal cables. The frequency response of the ECM-50 is 40 Hz to 14 kHz with a slight rise at the high end to allow for placement below the performer's chin. Although it is usually worn on a tie clip attached to a performer, it is also extremely useful as a piano pickup mike during solo classical performances where no visible mike is normally permitted. The ECM-50 weighs 0.3 ounce.

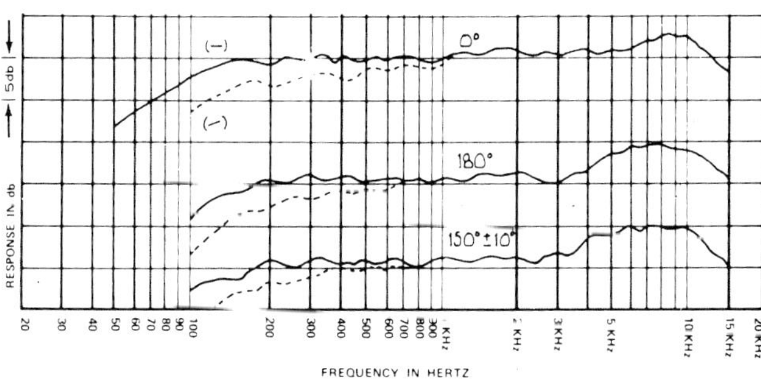

Figure 5–8C Frequency response of E-V RE-16. *(Provided courtesy of Electro-Voice.)*

Sennheiser MKH-816 Microphone

The MKH-816 is a shotgun mike with a club-shaped polar pattern (not supplied by the manufacturer) and a frequency response of 40 Hz to 20 kHz. It is less than an inch in diameter and 22.2 inches long. It weighs 375 grams. (See Fig. 5–9.)

Mike Cables

The cables that connect microphones to the mixing equipment are subject to the same abuse factors as the previously described camera cables, as are their connectors. We make the same plea here for care in their use. Unlike camera cables, often mike cables are unplugged when not in use. Save the embarrassment of a "si-lent" program by checking whether the cable connectors are plugged into their receptacles *before* the program or rehearsal starts.

Some Aesthetic Considerations in Microphone Use

On Mike and Off Mike.   A performer should be *"on mike,"* or within the polar pattern area of the mike, for normal pickup. *Off-mike areas* refer to the so-called dead sides, or areas, around velocity and cardioid mikes. These mikes are, of course, not totally dead to sound, but the pickup will have a far-away, down-in-a-barrel quality. Sometimes this off-mike effect is deliberately used to simulate aural distance between two performers.

Presence.   A performer's mike *presence* may be defined as his being on mike at *his* proper distance from the mike. If the performer is too close, then lip smacking, teeth clicking (particularly dentures), tongue slapping, and breathing noises will be heard. Then the performer is said to have "too much presence." If, however, the performer is too far from the mike, his voice will sound roomy, or hollow, and have a lackluster quality. Then he is then said to have "no presence."

Figure 5–9A The Sennheiser MKH-816 shotgun microphone. *(Provided courtesy of Sennheiser.)*

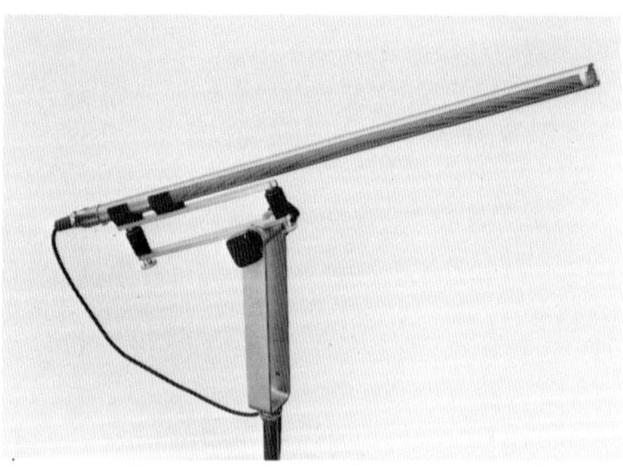

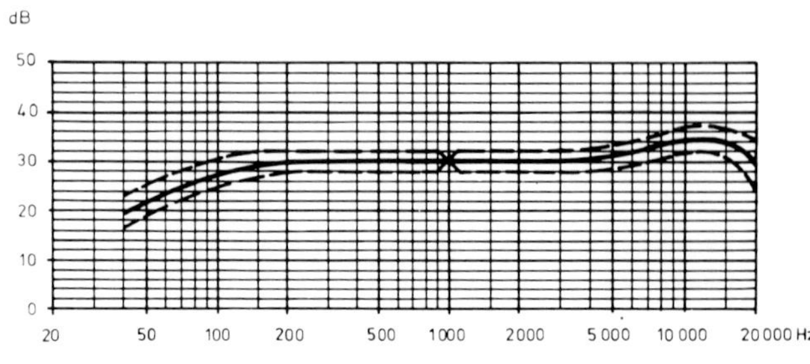

Figure 5–9B Frequency response of the Sennheiser MKH-816 shotgun microphone. *(Provided courtesy of Sennheiser.)*

Mike Phase Reversal. If there are two live mikes in the same studio that feed the same mixer from the same sound signal source, they must be in the same signal phase. If they are in opposing phase, or *out of phase,* then their combined signals will subtract rather than add. Recall that each mike has an internal transducer diaphragm that reacts to sound impinging upon it by creating a minute signal flow of alternating current. If one of the two mikes just mentioned had one mike cable whose "high" side wiring and "low" side wiring were oppositely wired from the other mike cable, then the signal currents flowing from the mikes into the mixer would be out of phase. The condition can be corrected by rewiring the reversed-polarity cable at its connector.

Multiple-Mike Interference. Frequency response loss can occur when two mikes with proper phasing are placed too close to each other. They should be placed so that there is at least a 3:1 ratio between mike and mike and between mike and user. That is, the mikes should be three times as far apart from each other as either is from the user. When two mikes *must* be placed close together, as on a podium, then multiple-mike interference can be avoided by placing the mike heads directly together. If the mikes are cardioids, the 3:1 ratio can be reduced somewhat by angling the mikes away from each other.

WIRELESS MICROPHONE SYSTEMS

Wireless mike systems use virtually every type of mike. So what we are discussing here is really not mikes, but radio-frequency transmission and reception systems to use with microphones instead of microphone cables. These systems have become very popular in the industry be-

cause of situations in which it is very difficult, or often impossible, to run mike cables. Examples are the news reporter on a crowded convention floor, attempting to interview a candidate; a performer on stage, singing while gyrating or rapidly moving across the stage; the performer singing on a barge stage, anchored 50 feet offshore, with the audience on shore; the speaker at the high school commencement, with the camera 500 feet away, on a long zoom. The solution to difficult or impossible mike cable runs is to use wireless mike systems. And the reason why these systems do not eliminate mike cables entirely is their cost. They are expensive.

In a typical system, the mike feeds into a miniature, battery-operated radio transmitter; or often the transmitter is integral to the mike case. The transmitter is very low powered, ranging from about 5 to 200 milliwatts (mW) for short-range transmission. The system's FCC transmission frequency allocations are in the 30 to 50 MHz very high-frequency (VHF), the 450 to 500 MHz, and 947 to 952 MHz ultrahigh-frequency (UHF) bands.

The transmissions are received nearby with receivers on the same frequencies, often with complex antenna systems to maximize reception, and the receiver signal output is fed to the audio mixer.

Comrex 450 RA/TA Wireless Mike System

The Comrex 450 system is small and lightweight (Fig. 5–10). The transmitter and receiver are each built into a similar case 3 inches by 5 inches by 1 inch. They each weigh 11 ounces. The transmitter easily fits into a performer's pocket. The controls on top of each unit are the on/off switch, the mike connector, and the

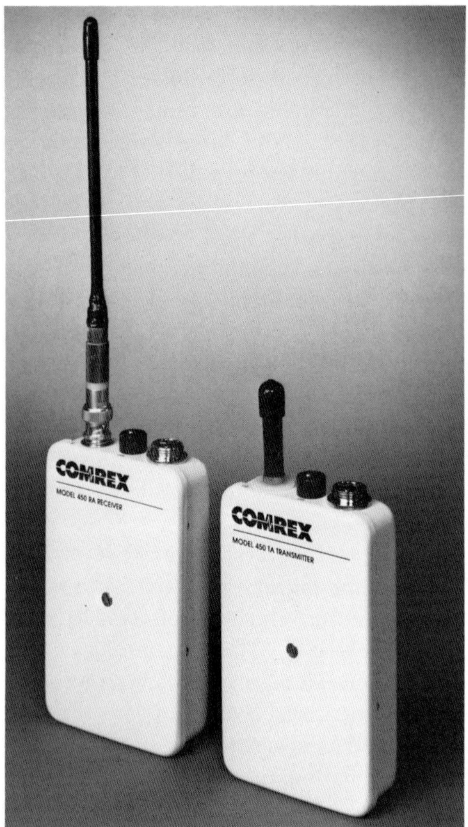

Figure 5–10 The Comrex 450 RA/TA wireless mike system. *(Provided courtesy of Comrex.)*

antenna. The transmitter operates on any two frequencies in either the 450 to 451 MHz or 455 to 456 MHz band, with a maximum frequency separation of 1 MHz. The power output of the transmitter is 150mW into a "rubber ducky" antenna; or it can feed a coaxial cable into a higher-gain antenna. Both transmitter and receiver are powered by two 9-volt (V) alkaline or nickel-cadmium batteries. Any low-impedance dynamic or condenser mike can be used, with condenser mikes being powered from the transmitter batteries.

Comrex HHT-1KA Microphone

The HHT-1KA is a self-contained dual-channel 1-W transmitter with a built-in electret mike (Fig. 5–11). It was designed for electronic news gathering (ENG) sound pickup. A mike/external switch and input jack permit the unit to be fed from a tape recorder or mixer output. Like the Comrex 450, the HHT-1KA operates on any two frequencies in the 450 to 451 or 455 to 456 MHz band. Its power source is six AA-type (penlight) alkaline or nickel-cadmium batteries. It is 10 inches long by 1.75 inches by 1.75 inches and weighs 14 ounces. The helical antenna mounts on its bottom.

The RRB receiver is rack-mounted and is compatible in frequency with both the HHT-1KA and 450 transmitters.

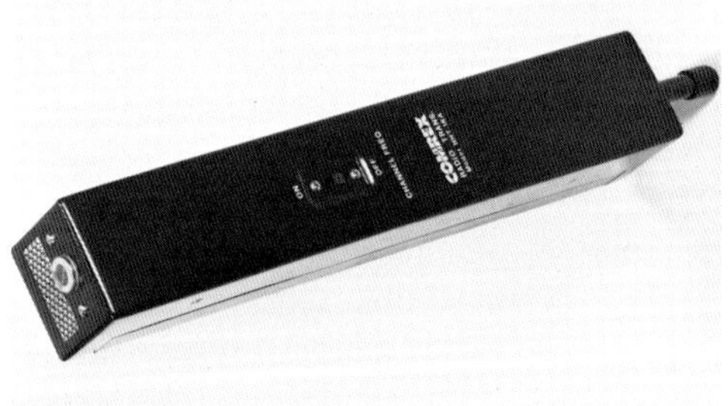

Figure 5–11A The Comrex HHT-1KA ENG mike. *(Provided courtesy of Comrex.)*

Figure 5–11B The Comrex RRB receiver. *(Provided courtesy of Comrex.)*

Micron CTR-101 Wireless System

The Micron CTR system consists of either the TX101 or TX102 transmitter (they are identical except in physical size and number of batteries) and the MR-1 receiver (Fig. 5–12). The system operates in the 30 to 500 MHz range, with the specific frequency chosen by the purchaser. It is a frequency-modulated (FM) system with a transmitter power output of 30 mW nominal. The transmitter uses an eight-pin mike connector, so that it can accept a transformer input from dynamic and condenser mikes, or a high-pass filter to avoid overload from wind noise and voice popping, phantom powering (simplex) for condenser mikes, and a 20-db attenuator for use with high-output mikes or high-output rock singers.

The receiver is externally powered from a battery pack, an alternating-current (ac) supply, or a recorder or mixer. The receiver has a unique A.R. (all-right) indicator which enables the operator to see at a glance that the transmitter is on frequency, that its batteries are in operable condition, and that the received signal strength is adequate.

As to signal strength, one of the weaknesses of a wireless system is that the signal often fades or disappears when the transmitter moves through signal *nulls.* To combat signal fade, Micron offers its MDS1 diversity receiving system. Two MR-1 receivers in differing locations, or in the same location with differently oriented antennas, are fed into the MDU-101 combining unit. The MR-1's are on the same frequency. The combining unit compares the signal strength of the two receivers. If the signal voltages are nearly equal, it combines the audio outputs of both receivers. If one of the signals is significantly weaker, the combining unit rejects the output of that receiver.

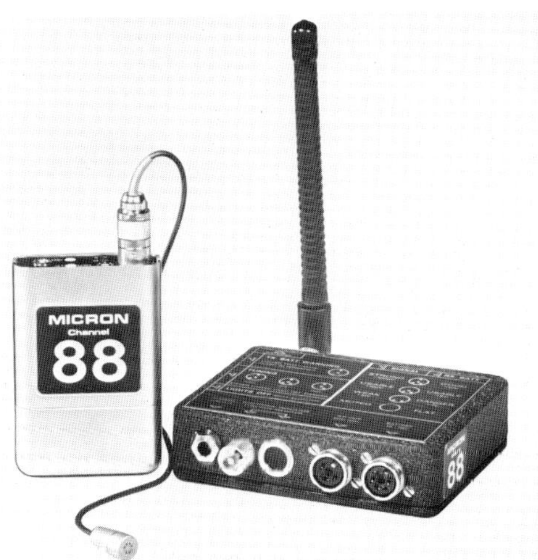

Figure 5–12 The Micron CTR-101 wireless system. *(Provided courtesy of Micron.)*

Sony WRT-27 and WRT-57 Wireless Mikes

The Sony wireless mike systems are capable of operating in the 900-MHz band, the 470-MHz band, or the 40-MHz band, or a combination of all three bands to maximize performance (Fig. 5–13). The system uses multiband operation to permit use of a great many wireless mikes during the same performance. Within an assigned frequency band, only a limited number of mikes can be used without causing interference problems. For instance, only five mikes can be used in the 900-MHz band and twelve in the 470-MHz band.

The WRT-57 mike/transmitter is a unidirectional condenser mike and an FM transmitter with 30 mW of power. It has a 6.25-inch wire pigtail antenna on its bottom. It uses a 9.45-V mercury battery. Its transmissions are received by either a

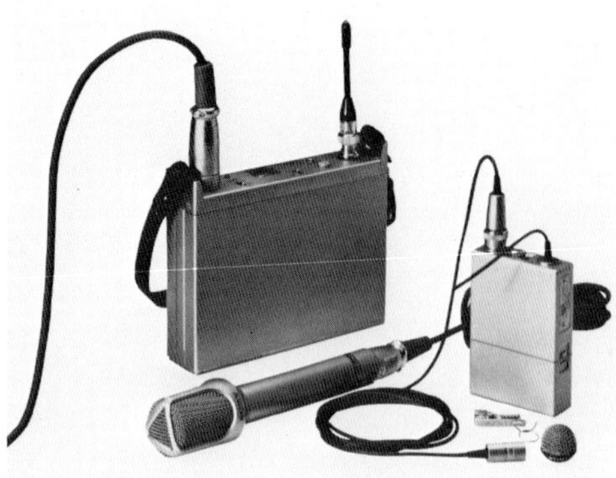

Figure 5–13A The Sony WRT-27 and WRR-27 wireless mike transmitter. *(Provided courtesy of Sony.)*

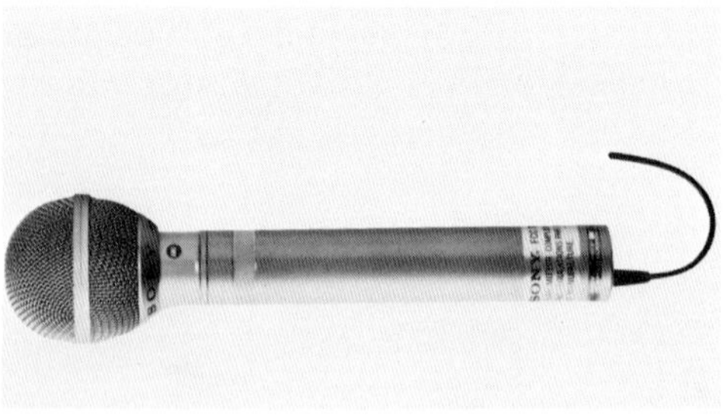

Figure 5–13B The Sony WRT-57. *(Provided courtesy of Sony.)*

ground-plane antenna or parabolic antenna which mounts on a microphone floor stand. The ground-plane antenna has zero gain, but the parabolic has 6 db of gain. Either antenna may be fed through an antenna booster amplifier which provides 18 to 20 db of gain in the UHF band (the 450- or 900-MHz band) to either a portable recever, the WRR-27, or to the WRR-57 (not shown).

The WRT-27 transmitter is a pocket model, and it is shown with two of the many mikes that can feed it. The WRT-27 is used exclusively in the 900-MHz band. It is powered by a small 9-V battery or by an external 12-V power supply through a power adapter which can be fitted to the microphone connector. The transmitter is FM and has a power output of 30 mW. It can be connected to a power amplifier, the WP-27 (not shown), to provide a power output of 500 mW to the antenna.

The WRR-55 diversity unit combines two antennas and two receiver/tuners to provide a significant decrease in signal dropout.

Figure 5–14 shows four unit combinations of the Sony wireless systems.

May the Mike Be Seen?

On a television program, mikes may or may not be permitted to be seen. This decision depends on the type of program being performed: panel show, yes; Shakespeare, no.

If mikes are permitted, current usage calls for tie-tack mikes like the Sony ECM-50. If the mike is not permitted in the picture, then boom-mounted unidirectional mikes are used since the mikes may be several feet away from the performer(s).

Microphone Stands and Mounts

Except for the tie-tack mike for which the performer *is* the mike holder, full-sized microphones employ one type of holder or another (Fig. 5–15). There are floor stands, in the heavy studio and folding portable variety, boom stands which project the mike above and over the user, desk stands in front of the user, and mike clamps for remote use on podiums. There are shock mounts which protect the mike from mechanical noise. And there is the stage mount, or "mike mouse," which is a foam plastic brick, rounded on top, flat on the bottom, grey in color, developed

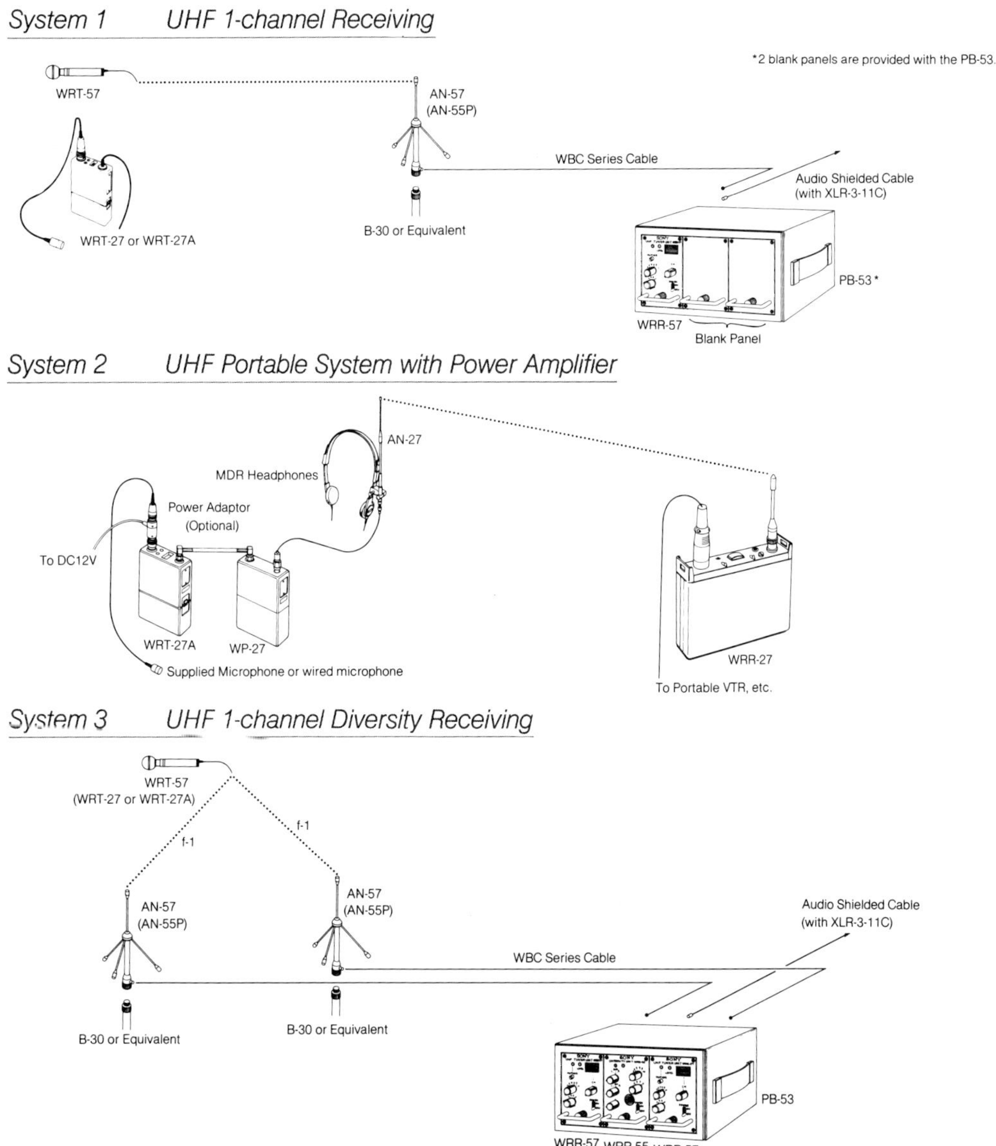

Figure 5–14A Sony wireless system 1, 2, and 3. *(Provided courtesy of Sony.)*

by Electro-Voice, into which the mike is nestled, with its cable emerging like a mouse tail. This mount sits directly on the stage, picking up reflected sound directly, without the delay normally associated with reflected sound.

MICROPHONE SETUPS

We discuss now the *placement* of micro phones for optimum sound pickup. We necessarily mention specific situations and conditions, but since these are sub-

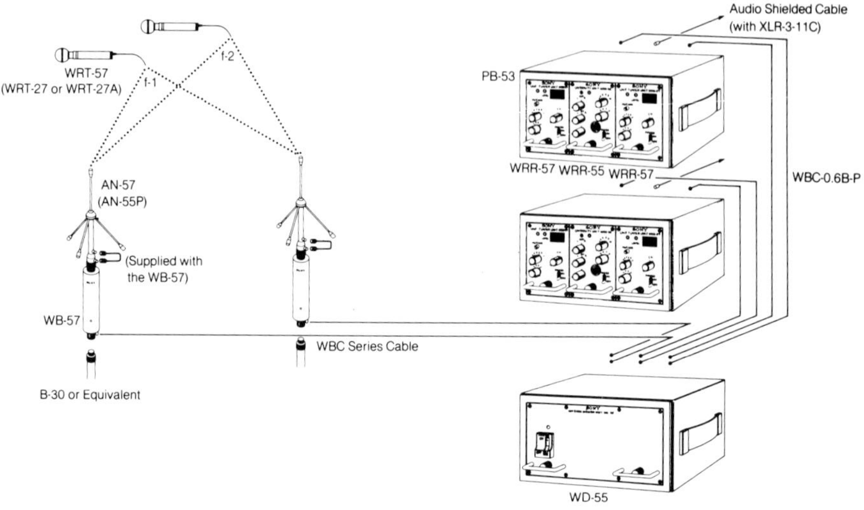

Figure 5–14B Sony wireless system 4. *(Provided courtesy of Sony.)*

ject to variation in real life, these setups must be considered as starting points rather than examples of cast-in-concrete rules.

First, let us look at setup complexity. Complexity is neither good or bad; it just *is*, and it has two aspects. The first aspect reminds the person doing the setup to "keep it simple." Never use two mikes where one will do. The fewer used, the less the likelihood of opening the wrong one. Consider the mike proximity effect, think about opposing phase conditions, and so forth.

The second aspect has to do with *what*, in terms of complexity, is necessary to get the optimum sound pickup. Sometimes it is necessary to use more than one mike to get the right pickup of even one musical instrument. Sometimes mikes must

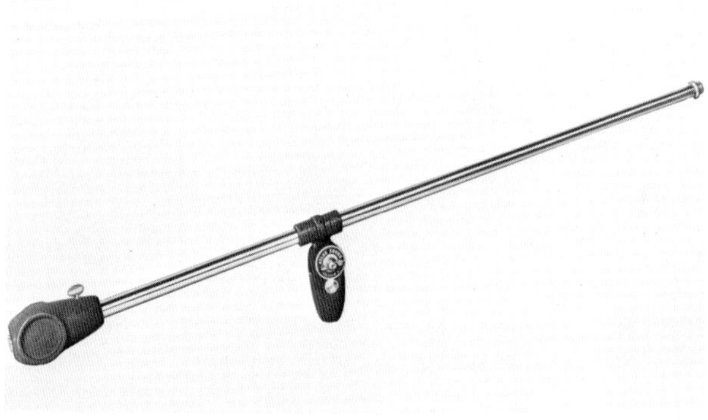

Figure 5–15 Mike stands and mounts. *(A–E). (Provided coutesy of Atlas and Electro Voice.)*

A

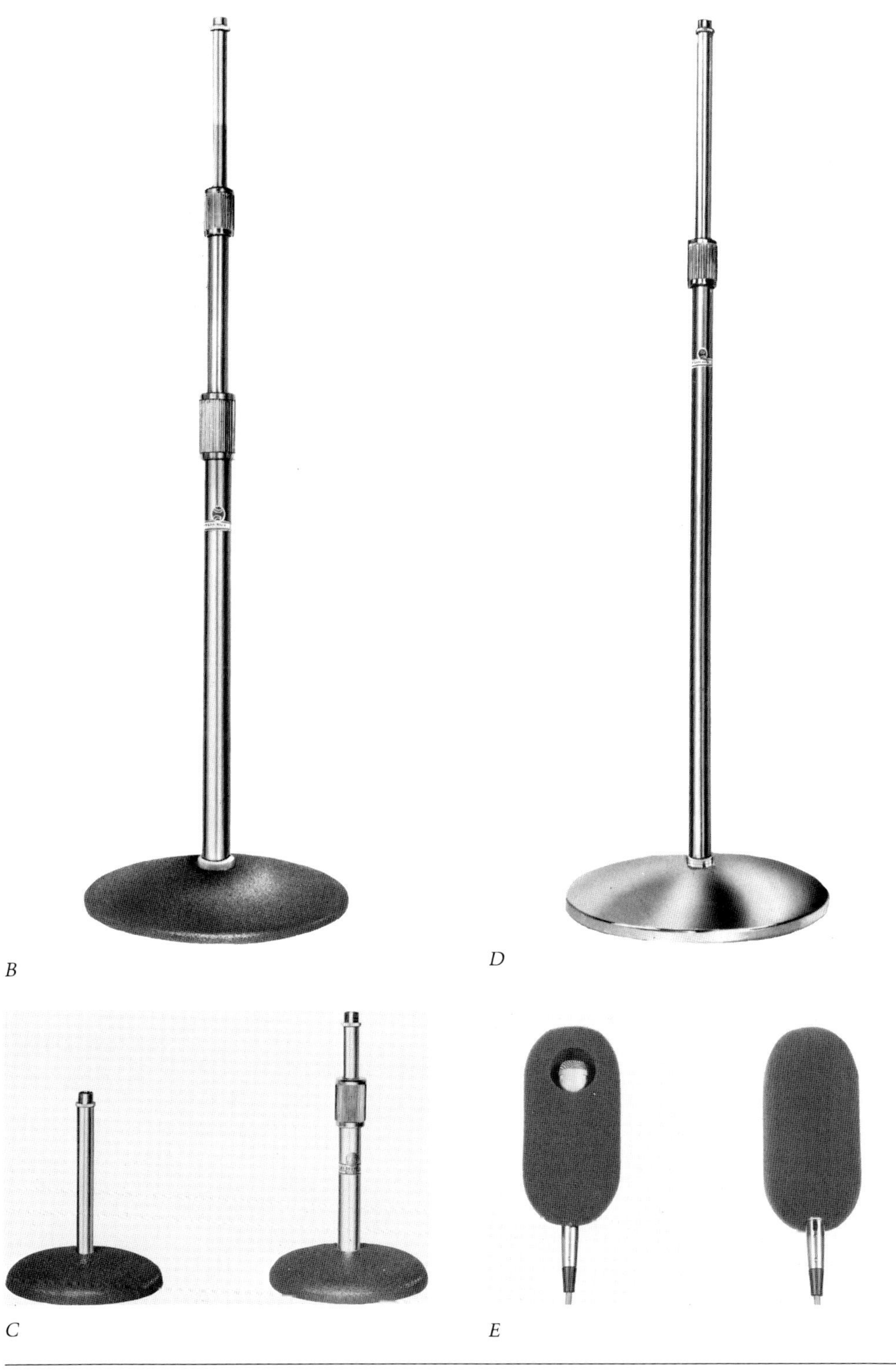

B

C

D

E

be angled *more* to avoid adjacent musical instrument pickup than to pick up the primary sound source.

Microphone arrangement, then, really amounts to microphone compromise. We start our look at that compromise by briefly reviewing what was stated earlier.

To get the desired sound, we must use a mike with the right directional characteristics and the right frequency response. The right pickup pattern is important as much to avoid unwanted sound pickup as to get the right sound. The right response curve is needed to include or preclude the bass proximity effect or high-frequency roll-off. A choice must be made, then, to use either a directional or an omnidirectional mike, and this decision should be made solely in terms of sound avoidance. If it is not necessary to use a directional mike to avoid spurious sound pickup, then use an omnidirectional one. Omnidirectional mikes are better mikes in terms of their avoidance of mechanical, breath, and wind noises, and they have a wider, broader frequency response than directional mikes.

Once the proper *type* has been selected—and it really is a difficult choice, because directional cardioid mikes are usually more expensive and one tends to use the most expensive tool available, equating expensive with best, to do a job that will receive critical judgment—the decision must be made as to placement of the mike. This decision, which will be refined as many times as is necessary, involves the distance between the mike and the sound source and, for the cardioid mike, the angle or cone of sound acceptance or sound avoidance.

Sound waves follow the inverse-square law: they decrease in intensity with the square of the distance between sound source and mike. So there is less sound available for pickup as the mike moves away from the source. Too, the farther away the mike is from the wanted sound source, the closer the mike gets to possible unwanted sources.

Start, therefore, with the mike close to the desired sound source. But remember that some sounds have multiple sources (musical instruments) and that some sources tend to move about (also some musical instruments, such as a jazz horn or reed).

Before the angle of a directional mike can be chosen, the characteristics of the mike and its polar pattern must be known, as well as the sound radiating characteristics of the source and the proximity and radiating characteristics of any unwanted pickup.

In terms of distance and angle, further considerations are whether the pickup is indoors or outdoors, whether the mikes may be in view, and whether the sound is that of a pop group or a symphony orchestra. An outdoor pickup will have less reflected sound, even with a bandshell reflector covering all or part of the music-making group. Outdoor pickups will require "tighter" miking. A symphony orchestra performing indoors produces a combined sound which is dependent on the reflection of the hall in which they perform. This requires miking that produces a balance of direct and reflected sound. Often this miking can be accomplished with a single mike or a single pair of stereo mikes hung in the right place, above and about ten to fifteen seat rows in front of the orchestra.

An example of unseen mikes is the television play in which display of microphones would compromise the context of the dramatic piece.

In terms of music pickup, the frequency response of the mike must always be wide enough to cover the range of the instrument being miked.

Speaking-Voice Pickup

Most often speaking voices are picked up with tie-tack mikes clipped to the performer. If larger mikes are used, place the mike in front of the user, with its height adjusted to about chin level and the speaker talking either directly into or at a 45° angle to the mike axis as a starting point.

Monaural-Music Pickups

For solo instrument pickup, a mike is placed at the optimum presence point for that instrument. This is determined in a general sense by the place on the instrument from which the *direct* sound emanates, for instance, the bell of a trumpet or the "S" hole of a violin. However, because its sound emanates from such a large area, the grand piano may be miked in one of three different ways. First, with the lid opened to about a 45° angle, the mike is placed close to the piano, perpendicular to the hammer line and aimed for the sound reflected from the lid. Second, a dynamic mike is aimed down into the third hole of the piano's metal bed, or frame. In this method, a tie-tack electret dynamic is used if the mike should not be seen. Third, two dynamic mikes are used, one located at the second hole (from the pianist) in the frame aimed at the keyboard, and at right angles to it, and the second mike located at the other side of the piano, midway between front and back, and aimed at a 45° angle to the keyboard. The mikes are fed to separate mixer inputs to achieve gain balance. The point here is that the degree of effort involved in miking solo instruments is proportional to the effect achieved.

Miking too close to string instruments will pick up rosin squeal and bowing noises; miking too close to woodwinds picks up reed and sputum noises. Miking a piano from underneath, from the back of the sounding board, will muffle all the high notes.

Mike placement must also take into consideration the arm movements of the musician; with a live audience, the mike must not hide the performer's face.

In a monaural pickup of a music group, the prime concern is the pickup of the orchestral blend as it would sound to a live audience. This is important; we mention it because mike placement itself can *color* the sound blend.

Miking a small music group of up to eight or ten instruments usually involves using three mikes, one at the front and center and one at each side, all aimed at (on axis to) the group. The center mike is opened first (hopefully in rehearsal); then the left and right mikes are opened for instrument balance. Less gain from the mixer will be required on the tympani side and more gain on the reeds. Consider movement of the side mikes or angling them for optimum music balance. During performance, solo instrumentalists are asked to step up to the center mike or to yet another mike which has been preset at the proper height for that instrument.

Miking a large orchestra is an expanded variation of the technique of small-group pickup. One central mike is used for overall pickup, but this time it is placed farther in front, so that its cone of sound accceptance encompasses the entire group. This is usually 15 to 20 feet away, on average, and 15 to 20 feet above the floor if possible, especially if the musicians are on risers. This latter mike requirement may necessitate hanging the central mike. The left- and right-side mikes may be augmented by separate mikes on particular sections or instruments in the orchestra, such as the strings or the woodwinds, because of the relatively low sound volume emanating from those instruments in comparison with that of horns and tympani. Instrumental and vocal soloists should be handled on a separate mike in front or to one side of the orchestra. This mike should be opened only as needed, and its gain balanced against the orchestra.

The response of any mike pickup will be affected by the acoustic parameters of the room, the type of construction, the angles of the walls and ceiling, the type of floor covering, and the absorptive effect of the audience.

How Many Mikes?

The only way to truly answer this question is in terms of minimums. A stereo pickup must have at least two mikes. A set of drums should have a separate mike with separate response for every vibrating surface (drumhead, cymbal). Amplified instruments should have a mike at the amplifier speaker, usually a variable-D cardioid, aimed midway between the edge of the cone and the center. A vocalist with acoustic guitar gets two mikes on the same mike stand, one at mouth level and the other at the height of the guitar hole and aimed at the hole.

Sound Avoidance

Before you set up any mikes, especially indoors, stop everything for a moment, with no one else around, and listen to the ambience, or room sound. Listen particularly for sounds that are not heard separately by the ears, because they are part of the background. Listen for fans (mechanical ones), air-conditioning rush,

creaking doors or floors. Clap your hands and listen for sound reflection, or echo. Remember that microphones, unlike ears, will not discriminate in favor of desired sound. Then, and only then, choose the mikes and position them to get the best compromise that good, patient audio practice will allow.

MIXERS, CONSOLES, AND CONTROL BOARDS

The terms *mixer, console,* and *audio control board* are almost synonymous. By common usage in the industry, a mixer is almost always portable and has five or less inputs, while a console or control board is usually fixed, in a control room or a production van, and may have many audio inputs, perhaps twenty to forty or even more. Remember, the number of inputs determines the name of the device only by common usage.

The inputs may be microphones; or they may be the outputs of other audio devices such as tape recorder playbacks and cartridge tape playbacks, record turntable playbacks, or perhaps line feeds from remote locations or from the audio output of a VTR.

Mixers or consoles are used to control the audio gain (level) upward or downward of its inputs, to blend or mix those incoming singals, to sometimes process the quality of that sound, and then finally to route the mixed sound to its destination(s). Consoles will always have a switch, either a pushbutton type or a key type, to turn on or off each input or alternately to *assign* its audio to a bus. Mixers usually do not have this feature. Both will have a *pot,* which is a knob or a vertical-slide fader, to raise or lower the gain of the input.

Both mixers and consoles have a volume unit (VU) meter for each *program* channel to visually monitor sound levels, and both have a master gain control to exercise total control of each output or program channel of the device. Consoles and control boards have an audio monitoring system with loudspeakers fed from a monitor amplifier. Mixers employ earphones which operate from the program amplifier. Consoles have a cuing and/or audition channel, feeding a separate speaker or earphones for presetting tape or disk starting points and levels, before they are mixed into the program.

Mixers tend to be lightweight and portable. Consoles for remote television tend to be heavier and transportable, and their input and output connections are made by plug connectors at the back of the console, rather than being hard-wired, as is normally done with larger broadcast studio consoles.

As mentioned earlier, a much more expanded discussion of consoles and control boards can be found in the author's earlier text, *Audio Control Handbook*, 5th ed. (Hastings House, Publishers, Inc., New York, 1983).

Shure M267 Mixer

We look first at a typical mixer, the Shure M267, which is perhaps the most popular mixer used in the industry (Fig. 5–16). The M267 weighs 5 pounds 2 ounces and has a frequency response of 30 Hz to 20 kHz. Its front panel has four rotary input pots, a master pot, and the VU meter. Above the input pots are low-cut filter switches for each input. The filters provide low-frequency roll-off to reduce the effects of wind noise or other low-end noise. Above the master pot is a *limiter*

Figure 5–16 The Shure M267 mixer, front and back views *(Provided courtesy of Shure Bros.)*

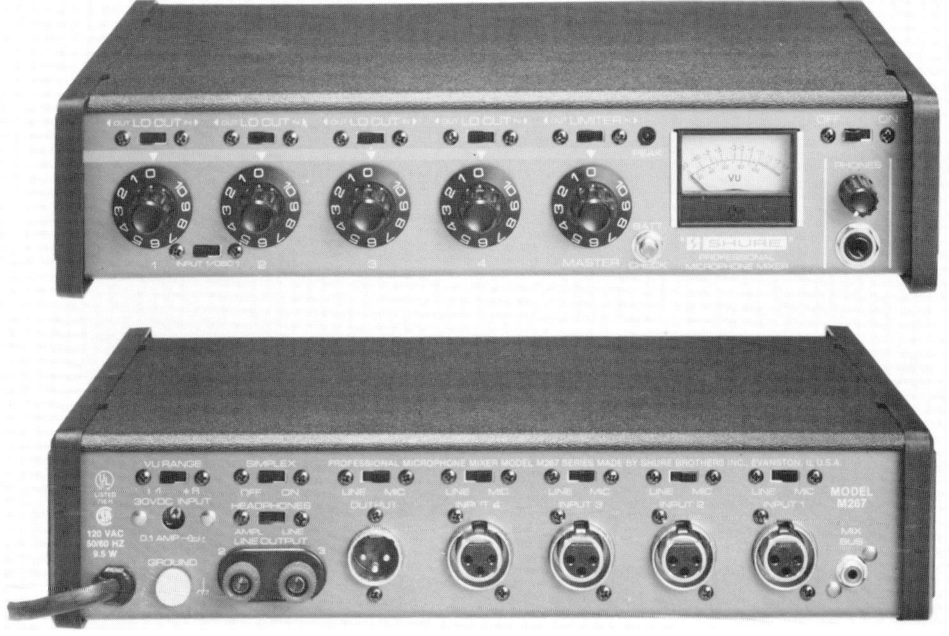

in/out switch which turns on a fast-acting, peak-responding, gain limitation circuit that cuts signal overload distortion during loud program intervals, without affecting normal program levels. The peak light-emitting diode (LED) indicator to the right of the limiter switch indicates limiter operation and flashes when program levels approach overload, with the limiter out of the circuit. The LED is much faster-acting than the VU meter and is activated by transient peaks.

A 1000-Hz tone oscillator is turned on and off by a switch between and below pots 1 and 2. Its level is controlled by pot 1, and its tone signal appears on line and mike outputs and on headphone and mix bus connectors. Thus the oscillator should be turned off when it is not in use.

The M267 power on/off switch is located to the upper right of the VU meter. In addition to ac power, the mixer can be operated from an internal battery supply consisting of three alkaline 9-V batteries. Access to the battery compartment is on the bottom of the mixer. With batteries in the compartment, the M267 will automatically and silently switch over to battery operation should the power fall below a suitable level or fail completely. If ac power fails, the operator is notified as the VU meter lamps go out. The battery condition can be determined by using the *battery check* (BATT CHECK) momentary switch located to the right and below the master pot, but only when the power cord is disconnected, the power switch is turned on, and the BATT CHECK switch is depressed. A new set of batteries will give about a +2 indication on the VU meter. The battery condition is still good if the reading is above 0 VU.

Battery life is about 20 hours of operation at +4 output in continuous use. Sporadic use lengthens battery life. If simplex powered mikes or high-level headphone monitoring is used, it will increase the drain on the battery supply and decrease battery life.

The rear panel of the M267 mixer has an output XLR connector, followed by four input XLRs. There is a LINE/MIKE switch above each connector which determines whether that input or output is set at mike or line level and impedance. Each input connector is directly behind the pot that controls the input. To the right of input 1 is the mix bus pin-jack connector, which permits *stacking*, or *ganging*, or *multing*, mixers for additional input capacity. With two M267s stacked, or multed, through this direct access to the mixing buses, the operator has two independent master gain controls, two isolated line amplifiers, and eight individually controlled inputs. Since the buses are directly paralleled, a 6-db gain drop will occur, and the master or input control level settings must be increased to compensate.

To the left of the XLR output connector is a pair of binding post connectors for line output, which is paralleled with the XLR and can be used simultaneously. The binding posts are marked no. 2 and no. 3, to correspond to pins 2 and 3 of the XLR, and are used in conjunction with the ground (earth) binding post to the left. When the M267 is used with ac power, its power supply is energized at the moment that the ac cord is connected to an ac receptacle (mains), and the unit should be grounded for safety.

Above the LINE OUT binding posts is the headphone amplifier line switch which permits choosing between the headphone amplifier (AMPL) and LINE, for operation when talk-back monitoring is needed on remotes. The headphone output connector appears on the front panel to the right of the VU meter, below the headphone level control. The headphone output level is high enough in the

AMPL (amplifier) mode to use as an additional unbalanced LINE output.

Above the headphone switch on the rear panel is a simplex (phantom power) switch to provide power to condenser mikes through the mike cable if they are used with the M267. Simplex power is not applied if the input mike/line switch is in the LINE position.

To the left of the simplex switch is a VU range switch which selects either a +4- or +8-dbm output at 0 VU meter indication. This switch changes the meter indication but does not change the actual output level. The meter is calibrated for use with a 600-ohm (Ω) terminated line. The +4 range is recommended for normal use to provide approximately 14 db of headroom from operating level to clipping level. The +8 range should be used with the limiter because limiting action begins at +7 VU.

Below the VU range switch is a 30-V externally fused power plug input for externally powering the M267 from a low-voltage source.

TEAC Model 30 Mixer

The TEAC model 30 (Fig. 5–17) replaces the TEAC model 3 mixer, which is a very popular audio control board in cable television. Note that although the manufacturer calls the model 30 a mixer, we refer to it as a control board, or perhaps a consolette, to conform to our earlier definition of the difference between a mixer and a control board.

The model 30 is a modular control board with eight inputs. The term *modular* indicates that each module, or section, is a complete entity which can be unplugged, removed, and replaced in the console, greatly expediting repair to an ailing console.

Input channels 1 through 6 are identical. Input channels 7 and 8 are different from the first six and are also identical. We describe each type of input channel.

There are two additional phono stereo inputs—an eight-input stereo submixer facility and four monitor inputs with pan-pot controls—to a stereo monitor channel. There are four VU meters, with each incorporating an LED peak reader. Two of the meters are permanently assigned to line outputs 1 and 2; the other two are switchable to buses 3 and 4, to the monitor channels, or to the submixer channels.

Interconnecting of separate console entities, where necessary, is performed on the back of the TEAC 30 by using jumper cables with RCA pin jacks. Only the mike inputs use XLR connectors.

In Fig. 5–18, from top to bottom, the first control is a mike attenuation (MIC ATT) three-position switch. This provides a 0-, 20-, or 40-db loss, in part by a pad and in part by changing the gain of the mike preamplifier, to prevent very loud sound from overloading the module preamplifier. The second control is another three-position switch. The first position sets the input at a low-impedance (−60 db, 150 to 600 Ω) mike level. The second position REMIX, selects the TAPE IN connector on the back of the module. This special input can be assigned simultaneously to the input and the submixer section, so that TAPE IN can feed the submixer and a mike can feed the input. The third position on the input selector is a standard line level input which selects the LINE IN jack on the back of the module. Since there are two high-level jacks on each input module, a total of 16 lines can be connected and routed without patching.

Below the two switches is a three-control parametric (sweep-type) equalizer. In

Figure 5–17 The Teac model 30 mixer. *(Provided courtesy of TEAC.)*

the circuitry before the equalizer there are patch points accessible on the rear of the module, to enable insertion of other external signal-processing devices. The equalizer consists of a 12.5-kHz shelving section, a 1- to 10-kHz section, and a 60-Hz to 1.5-kHz section.

The equalizer is followed *in the circuit* by the *input fader* control and another stage of amplification, in turn followed by a DIRECT OUT output at the back of the module. This is followed on the module panel by the MUTE switch, which lifts the module from the module assignment facility and feeds it so it can be used as a reverse solo system in REMIX. By releasing the mute switch a single signal can be heard without shutting down the

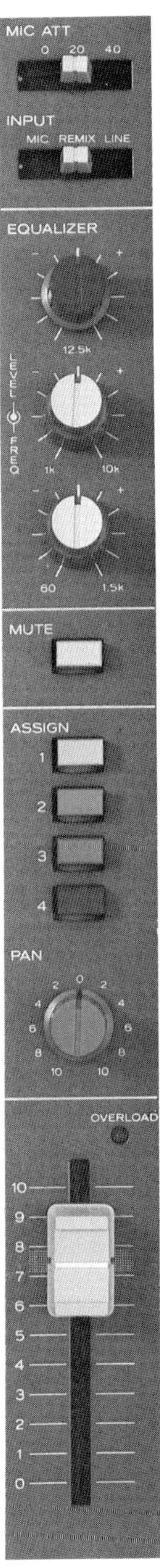

Figure 5–18 Teac 30 Input module. *(Provided courtesy of TEAC.)*

other fader settings. The mute switch may also be used to add a signal to a mix without fading it in.

The *assignment* switches below the mute are color-coded and numbered to coincide with the buses. They assign the input module to any of or all the output buses, complete with panpot facility between any two buses.

The submixer assignment of each input module follows the mute switch in the circuit, but is located physically on a submixer panel section at the extreme right of the console.

Input channels 7 and 8 are similar to the first six input channels except that the mike input structure is different. The mike input connectors are the PL 55 type of phonograph jacks rather than XLRs; there are no input transformers, so the inputs are unbalanced and they are high-impedance inputs. These input channels can be used to feed direct, transformerless input instrument mikes (on guitars and basses, for instance) to the mix, or a conversion kit can be added to the console, to make input channels 7 and 8 identical to the first six channels.

Each of the eight submixer controls consists of a three-position switch and a dual concentric pot (Fig. 5–19). The switch provides input from PRE, POST, and TAPE on the input module. PRE is from the input module, previous to any signal processing, just before the *access send* point. POST follows signal processing and is picked up in the circuit between the mute switch and the ASSIGN panpot; and TAPE is a direct feed from the TAPE IN connector on the back of the module.

The dual concentric pot for each input to the submixer provides gain control on the upper section and panpot on the lower section of the control. The submixer master pot below provides master gain control to the entire submixer.

Figure 5–19 The submixer module. *(Provided courtesy of TEAC.)*

The next section of the console to be described is the monitor section, located to the right of the input modules (Fig. 5–20). From top to bottom, the controls are the meter switch, which connects meters 3 and 4 to buses 3 and 4, to left and right monitor channels, or to submixer left and right outputs.

Below are the headphone gain control and headphone switch. The headphones can be selected to the monitor, the submixer, or to a center OFF position.

There are four concentric gain/pan monitor controls, one each for each of the four bus outputs, and below them is a master monitor rotary control.

Below the monitor controls is the console master program channel gain control, a vertical control like the input channel gain controls.

Finally we want to look at the back of the TEAC 30 (Fig. 5–21). Note that the module numbering is from right to left, to conform to the fact that the module backs are facing the viewer.

The bottom connectors on modules 1 to 6 are mike connectors, XLR females, and on modules 7 and 8 are phonograph jacks, mike connectors for high-impedance, unbalanced inputs. All the other connectors are RCA pin jacks.

At the top of each module, 1 through 8, are the access connectors for external signal-processing devices. Below them are the TAPE and LINE IN connectors. Below these are the DIRECT OUT and CUE OUT connectors. CUE OUT is an output feed to an external cuing system, taken directly from the PRE point on the PRE/ POST/TAPE switch on the submixer module.

To the left of the input modules are the two PHONO IN (left and right) and two PHONO OUT (left and right) facilities. They are followed by the four bus facili-

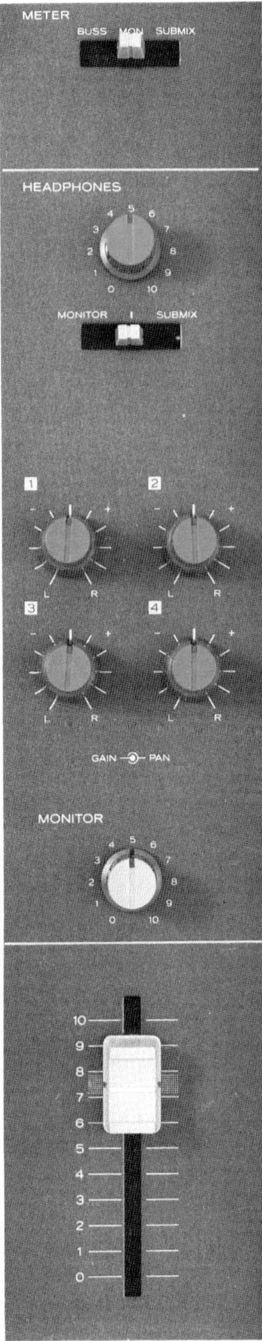

Figure 5–20 The monitor module. *(Provided courtesy of TEAC.)*

ties, with access connectors for external ECHO IN and OUT and BUS IN / LINE OUT connectors.

Below the bus inputs and line outputs are SUBMIXER IN and OUT (left and right), MONITOR OUT (left and right), and the power cord.

See Fig. 5–22.

The following data are taken from the specifications sheet for typical performance:

- Frequency response from any line input to any line output bus or monitor output or submixer output—30 Hz to 20 kHz, ±2 db
- Equivalent input noise: −113 db unweighted, 20 Hz to 30 kHz
- *S/N* ratio:
    8 lines to 1 bus − 64 db unweighted
    8 mikes to 1 bus − 53 db unweighted
    Bus to line 60 db at 1 kHz
    Input switch, 1 kHz, 65 db at nominal level
- Total harmonic distortion (THD):
    Mike input to bus output, 0.07% — Both at 1 kHz, nominal level
    Line input to bus output, 0.07%

Ward-Beck Systems Ltd. T-1202 Transportable Console

The WBS T 1202 weighs 120 pounds and would be used only at a remote location when a very high audio quality program were being contemplated. The term *transportable* refers to the fact that all inputs to, and outputs from, the console appear on the console back on XLR connectors rather than a hard-wired terminal strip, as is found in permanent console installations. (See Fig. 5–23.)

Figure 5–21 The TEAC model 30 back view. *(Provided courtesy of TEAC.)*

Note that all the inputs are female XLRs and all the outputs are male XLR connectors, which is the industry standard for audio wiring.

The WBS T 1202 is a modular console utilizing 12 standard input modules (designated the M480B module), two M487 master modules, and two M482B monitor modules (Fig. 5–25). The console's facilities include 12 input channels, each with a three-section equalizer and a separate Penny & Giles vertical fader made in the United Kingdom; two matrices of six-switch line input selectors, each capable of selecting among OFF, program bus M1, and program bus M2 positions; an auxiliary mixing bus assignable from each input channel and a cue mixing bus assignable from each input channel and each input of the line input selector; two monitor channels with 13 selectable sources; an internal test oscillator; and two VU meters, one assigned to M1 and the other selectable to M2, AUX, MON 1, MON 2, and OSCILLATOR.

M480B Input Module

Each input module (Fig. 5–26) provides the following facilities: At the top of the module, A or B input source (choice of one), with mike or line level selection and

rotary sensitivity control over a 30-db range applicable to either mike or line inputs; LED peak indicator to the right of the mike/line button; below that the PRE or POST AUX (auxiliary) switch, which allows the pickup point for the auxiliary signal to be before or after the channel fader, then the AUX switch which, when depressed, feeds the signal to the auxiliary mixing bus, together with the auxiliary level control. Below the auxiliary controls are the equalizer controls and equalizer insert switch. When depressed, this switch inserts the equalizer section into the signal path and activates the LED indicator. When the switch is released, the unequalized input signal is bypassed to the output amplifier. The CUE switch follows, which feeds the signal originating ahead of the fader to the external cue/solo mixing bus. At the bottom of the module are the two program bus assignment switches, M1 and M2.

Two master output channels are available on this console. Each provides two isolated sources; one is brought to an XLR connector, and the other is used as a source for the two three-button line output switches. Each master program channel utilizes an M487 master module and a Penny & Giles master vertical fader. The two master program channels provide the following facilities: Assignment in any

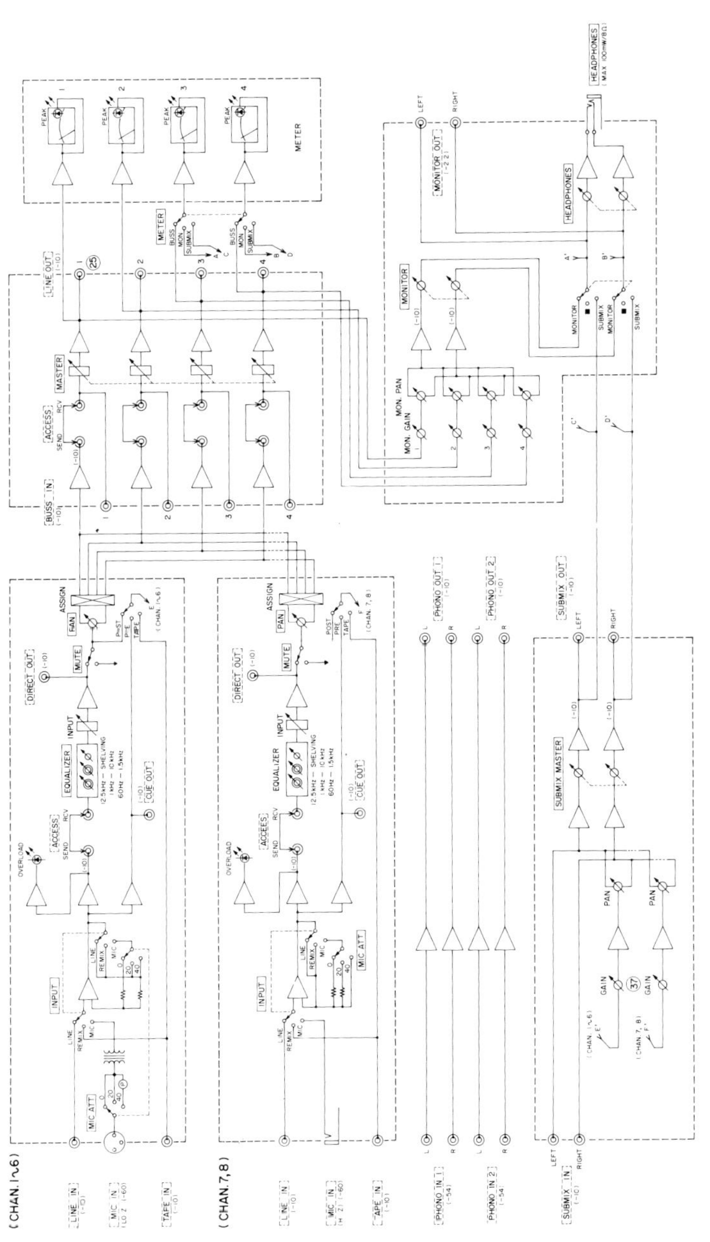

Figure 5–22 Teac 30 functional diagram. *(Provided courtesy of TEAC.)*

Figure 5–23 Ward-Beck T 1202 console. *(Provided courtesy of Ward-Beck.)*

combination of the 12 mike/line input channels; direct inputs, permitting multiconsole interconnection; master 2 to master 1 reentry, with reentry ON pushbutton and rotary level control; 1-kHz tone on the master output lines for setup purposes; and VU metering with one fixed meter for M1 and one switchable meter for M2.

One auxiliary output channel is provided and includes the following facilities: Assignment from each of the 12 input channels via an on/off pushbutton switch, PRE/POST pushbutton switch, and a rotary level control; a rotary auxiliary master level control; auxiliary output is terminated on an XLR connector, and VU monitoring is via the switchable meter.

Two monitor output channels are provided as well as a cue bus with prefader listening (PFL) and solo capabilities (Fig. 5–27). The two monitor channels each consist of an M482B monitor module, and each provides the following facilities: Pushbutton selection among the following sources: Six line selector inputs, M1 bus, M2 bus, auxiliary bus, cue bus, UTIL

A, B, and C inputs; rotary monitor level control; separate direct and switched outputs, each terminating in an XLR connector. The direct output is suitable for feeding utility output lines or headsets. A seven-pin XLR headset connector is also provided. The switched output is suitable for feeding an external monitor amplifier; a solo facility whereby the cue channel, when in the solo operation mode, will interrupt the selected monitor source for as long as a CUE pushbutton is depressed; and VU monitoring via the switchable meter.

A cue channel is provided which is selectable from any of the 12 console input channels or from any of the inputs to the six-position line input switcher. The cue channel may be operated in either of two modes: PFL or solo. When operation is in the PFL mode, activation of any CUE pushbutton will cause the selected channel(s) to appear on a built-in cue speaker. The signal level for this speaker is controlled by a rotary level control. When operation is in the solo mode, activation of any CUE pushbutton will cause the selected channel to appear on the console

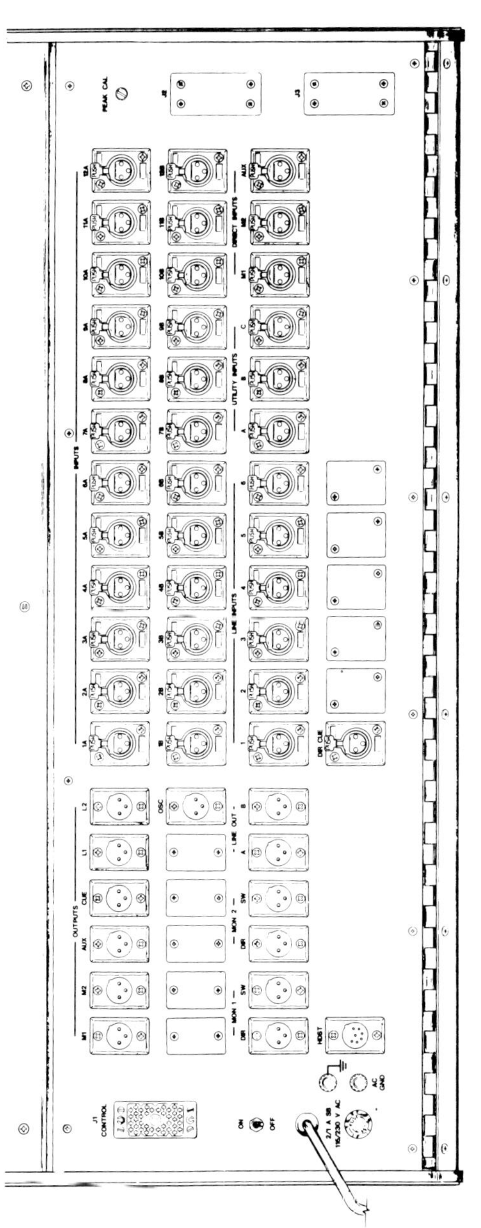

Figure 5–24 WBS T 1202 back view drawing. *(Provided courtesy of Ward-Beck.)*

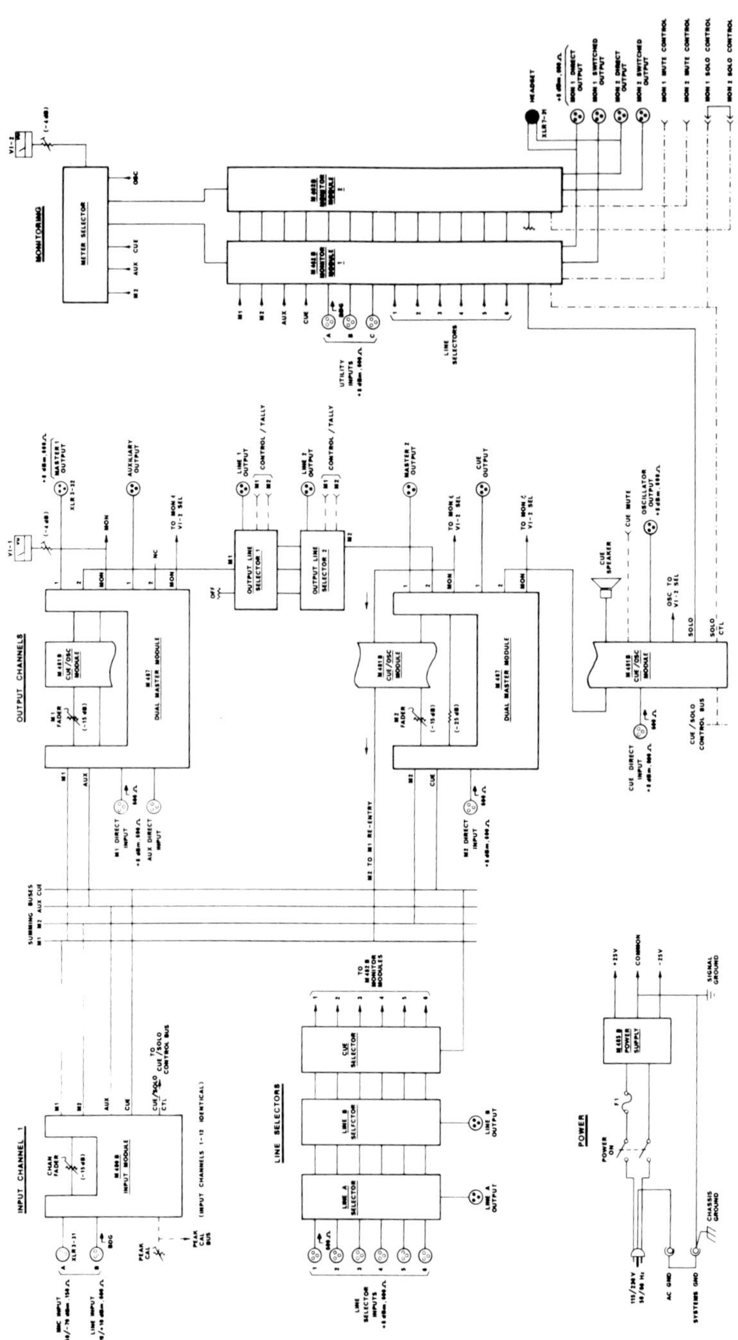

Figure 5–25 Functional diagram drawing. *(Provided courtesy of Ward-Beck.)*

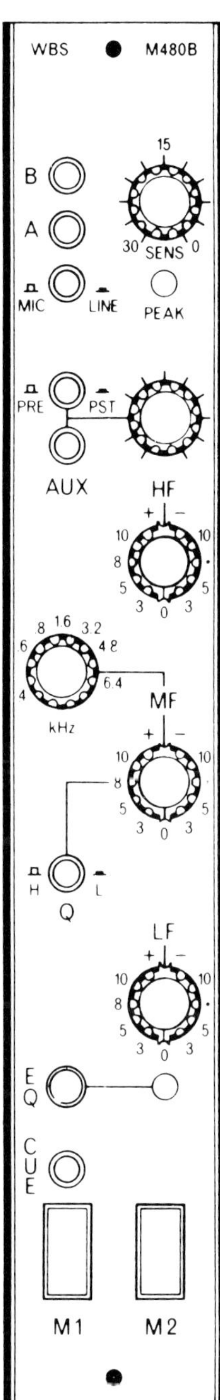

Figure 5–26 Input module drawing.
(Provided courtesy of Ward-Beck.)

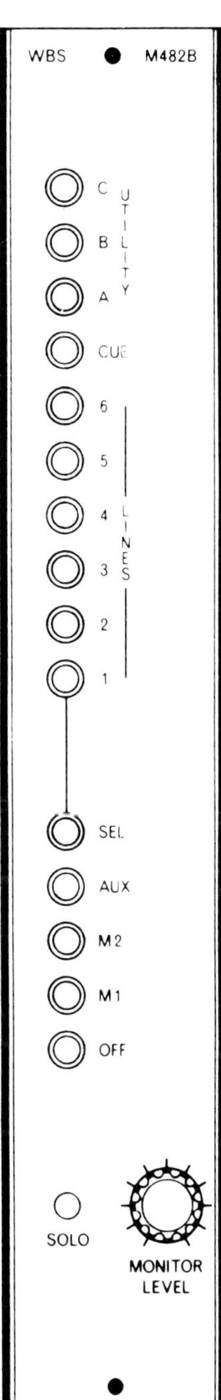

Figure 5–27 The monitor module drawing.
(Provided courtesy of Ward-Beck.)

monitor output, interrupting the signal normally heard. The cue channel has the further capability of acting as a second auxiliary channel, with its output appearing on an XLR connector and on the switchable meter selector.

Two six-position line input selector switches are provided, with input and output connections on XLR connectors. They are used by connecting their outputs to selected input channel line inputs. They have interlocking switching, and their inputs are also selectable to the cue bus.

Two three-position line output selector switches are provided. Sources for these selectors are the M1 and M2 output buses, and control logic from these selectors is provided for interlock and muting purposes.

The cue, oscillator, reentry module is located to the left of the monitor modules (Fig. 5–28). It contains, in descending order, the auxiliary master level control, the reference oscillator switch with screwdriver trim control below it, the solo level control, the cue/solo assignment switch, the cue level rotary control, and the M2 to M1 reentry switch with rotary reentry level control beside it.

From the specifications sheet come the following data:

1. *Inputs*

    Mike, greater than 5 times source Z, balanced and floating.

    Headroom: 24 db for all input sensitivity settings.

    Level: Sensitivity continuously adjustable over a 30-db range from −70 to −40 dbm.

    Gain: 78 db nominal, 108 db maximum, transducer gain.

    Line: Input Z; 30 kΩ balanced.

    Headroom: 24 db for all sensitivity settings.

    Level: Sensitivity continuously ad-

Figure 5–28 The cue, oscillator, re-entry module. *(Provided courtesy of Ward-Beck.)*

justable over a 30-db range from
−20 to +10 dbm.
    Gain: 28 db nominal, 58 db maximum, bridging gain.
    Line input selector and direct inputs:
    Source Z: 600 Ω nominal.
    Input Z: 600 Ω balanced.
    Input headroom: 16 db above nominal operating level.

2. *Program Outputs*
    Output Z: 600 Ω balanced.
    Load Z: 600 Ω balanced.
    Nominal output level: +8 dbm
    Maximum output level: +24 dbm

3. *Frequency response:* ± 1 db from 30 Hz to 20 kHz.

4. *Noise*
    Mike, −125 EIN over a bandwidth of 20 to 20 kHz.
    Line, 75-db S/N ratio, reference +8 dbm output level over a bandwidth of 20 to 20 kHz.

5. *THD:* 0.25% at 10 db above nominal operating level, from any input channel to any program output channel, 30 Hz to 20 kHz.

6. *Crosstalk:* Signal-to-crosstalk ratio greater than 70 db from 30 Hz to 15 kHz, measured on any signal path through the console with signal fed to any other console signal path.

Audio Tape Recorders and Cartridge Machines in a TV Control Room

Primarily to include music in the form of themes, music bridges between scenes, and background music to fill the silence gaps and dramatic pauses, we use audio tape playbacks in the TV control room. Two distinct types of playbacks are found: the open-reel type and the cartridge type. Cartridges are used primarily for short music pieces.

An open-reel (reel-to-reel) playback will have a speed and equalization control that will put the machine in the 3.75, 7.5, or 15 inches per second (ips) speed; and it

will have a mode switch which can be positioned to play, record, fast-forward, fast-rewind, or stop. The tape heads are aligned between the feed and takeup reels in this order: erase head, record head, playback head. The azimuth settings of the record and playback heads are critical and occasionally must be readjusted. (See Fig. 5–29.)

Audio tape playback is at 3.75 ips or at 7.5 ips in the television control room. The higher speeds of 15 and 30 ips are used by the audio recording industry for maximum ease in editing and highest quality of reproduction. Reproductive quality in audio recording is a function of the speed of the tape, the width of the track, and the width of the head gap. Audio heads remain fixed; they do not spin as video heads do. Quality, or frequency response, and the signal-to-noise ratio increase as the tape speed is increased, as the head gap (the space between the pole pieces of the head electromagnet) is narrowed, or as the tape track is widened.

The track is the magnetic signal that is "layed down," or written, on the tape by the record head, in a horizontal direction, in the tape's direction of travel. A monaural recording has a single track that is virtually the full width of the tape. A multitrack recording uses a small portion of the width of the tape for each track, with an unrecorded space between tracks. The width of each track is dependent on the number of tracks recorded on a given tape width.

Figure 5–29 Tape path drawing.

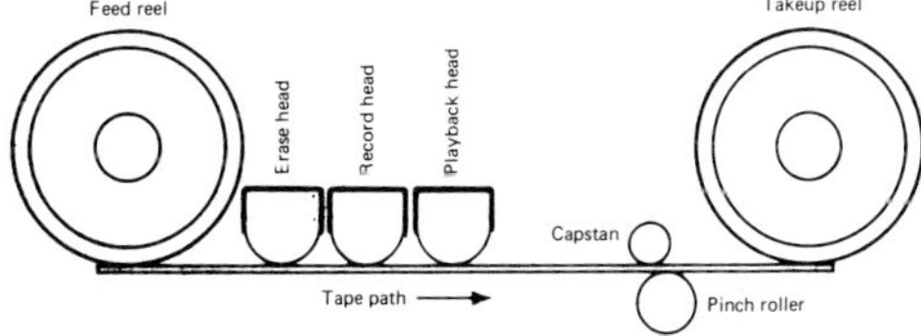

As the number of tracks is increased, the *noise-to-signal ratio* also is increased. The noise is that which is inherent in both the tape and its amplifiers. The noise level increases because, with a greater number of tracks on a given width of tape, each track becomes narrower, requiring greater amplification and hence increased inherent noise.

The recording level on audio tape is measured in terms of its magnetic properties and is stated in nanowebers per meter.

Tape Machine Operation

To operate any open-reel audio tape machine, place a reel of tape on the feed spindle (usually on the left side) and secure the reel. Pull the tape end through its tape path around the tape tension idlers, across the heads, between the capstan and pinch roller; and attach the tape to the takeup reel. Switch the machine to the desired speed. To play a recorded tape, switch the machine to the play mode, start the machine, and open the output pot (if there is one), feeding the modulation to an input of the audio mixer.

Cartridge Tape Reproducers

Cartridge machines, usually called "cart" machines, employ a continuous unending loop of specially lubricated tape in a plastic container. Tape cartridges have virtually replaced record turntables for playback of short music inserts. They have a unique advantage over turntables—instant cue-up. A cue tone burst is recorded on the cart tape on a separate cue track immediately before the start of program modulation. After a cart has been played, the tape continues to roll silently through the cart until the cue tone is reached, which instantly stops the tape motion. Then the cart is tightly cued and ready for the next play, even if it is removed from the machine. It remains cued until its next use.

One word of caution. Most cart recorders have no erase head, so carts must be bulk-erased before they are used again.

TV Studio Intercom Systems

Earlier we noted that an intercom earphone jack and level control were found on all the cameras we discussed. Camera intercom is one part of a small network of intercommunication between a television studio and its control room.

The television control room is kept in a semidarkened state, the better to view the monitor pictures. The control room may or may not be directly adjacent to the studio under its control. In a radio control room, a double-glass tilted window separates it from its studio, and they are both well lighted. Communication between these rooms exists through an *auditioned* mike when no other mikes are live; or it exists through a set of well-known hand signals among cast, crew, and director, through the double-glass window. The standard hand signals are pictured and described in this author's *Audio Control Handbook*, 5th ed.

In a television control room, the director must be in constant communication with the camera operators. Although most programs start out with camera shots throughout the program being carefully blocked out and entered on the operator's shot list, changes are always necessary. The director, therefore, wears a headset with one earphone and a talk-back mike attached, very much like a telephone operator's headset. This headset plugs into a jack in the director's table in front of him. The director must also be in contact with cast members—to cue them to start or stop action or dialog, to speed their pace or slow it down. But the cast member(s) cannot always wear an earphone, or listen while speaking, so the director speaks to a floor director, or floor manager, who wears a headphone and talk-back mike. The floor manager, standing next to the cameras but out of camera range, relays the director's cues, using the same hand signals as in radio. The director switches his talk-back mike and earphone from camera operators to floor manager to other required stations, with a set of illuminated pushbutton switches on the front of his desk.

On a network level program, individually manned mike booms are used. The boom operators must be directed to their next action area of the studio, and the audio operator in the control room performs this task. Both the audio person in the control room and the boom operators are connected for communication by an earphone/mike headset. The boom operators wear two earphones: one feeds them instructions, and the other is connected across the program line so that they can hear program audio.

Interestingly, the two most troublesome parts of a television production system are the intercom subsystem and the audio subsystem. Neglect of either will cause havoc in a control room.

REVIEW QUESTIONS

1. What is a transducer?
2. What are the components of the sound wave?
3. What is a polar pattern? A response curve?
4. Describe the peculiarity factors of a dynamic, a velocity, and a cardioid microphone.
5. What is presence?
6. Describe mike phase reversal.
7. Why are wireless mikes used?
8. Discuss mike setup in terms of sound avoidance.
9. What is the difference between a mixer and a console?
10. Describe an input module.
11. Why is audio tape used in television?
12. Describe television intercom systems. Why are they used?

Chapter 6

Equipment Interconnection, Impedance, and Other Technical Considerations, Including Trouble Diagnostics

This chapter is for the technically concerned, as differentiated from the solely operational, user of television equipment. There are some video parts and some audio parts, and others that fit both categories. We look at the video parts first.

VIDEO EQUIPMENT INTERCONNECTION

To interconnect the various pieces of equipment found in a video system—the camera, switcher, VTR, camera control units (CCUs), time-base correctors, and monitors—we use cables and connectors. Two types of cable are used to connect devices: that which carries the video signal and that which carries both signal voltages peripheral to video but at lower frequencies and powering voltages such as tally light power, logic control, or studio on-air light warning power. All circuits that do not carry video signal employ thin, stranded wire pairs, but the video signal is transmitted along *coaxial cable* because of its high frequency and low voltage, which is normally 1 V peak to peak. That is, from a zero voltage baseline, the video signal should be 0.5 volt (V) in the positive direction and 0.5 V in the negative direction.

Coaxial cable shields the video signal from intrusion by radio-frequency (RF) fields, such as citizens' band (CB) radio transmissions and other unwanted signal, for example, noise or power line hum. The coaxial cable most used in cable TV studio "runs" is designated RG-59U, is about 0.25 inch in diameter, is semiflexible, and has a nominal impedance of 75 ohms (Ω). We discuss impedance later. The reason that coaxial cable is not more flexible is that it is made by surrounding one solid copper wire or multistrand wire, of a specific diameter, by a semisolid plastic spacer of specific diameter, which is in turn surrounded by a copper braid or aluminum foil and covered by a plastic abrasion-resistant jacket. The distance between the center conductor and the braid or foil must be constantly maintained, all along the cable, to keep the impedance at 75 Ω. No tight turns or bends should be made in the cable which would degrade the coaxial cable. There is even a triaxial cable, which permits extremely long cable runs without degradation of the video signal.

In a multipin connector cable of 20 or 33 pins, such as those used between the editing controller and VTR or the 30-pin connector between the camera and CCU, at least two of the pin pairs and two elements of the cable are coaxial, while the rest of the cable is composed of stranded pairs. The two coaxial segments carry video from the camera to the VTR or CCU and return the monitor video to the viewfinder.

When coaxial cable is used alone, it is connected with F-type screw-on connectors, where the center pin of the connector is the coaxial center wire, with the BNC (a bayonet quarter-turn type), or with PL-259 professional connectors. The F type is the least expensive, and the PL-259 the most costly. See Fig. 6–1.

Figure 6–1A Video connector PL 259. *(Provided courtesy of Amphenol.)*

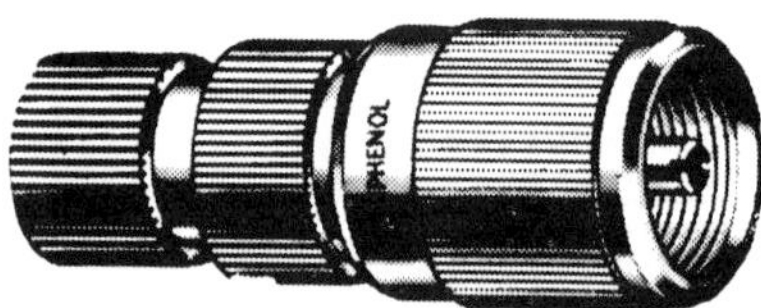

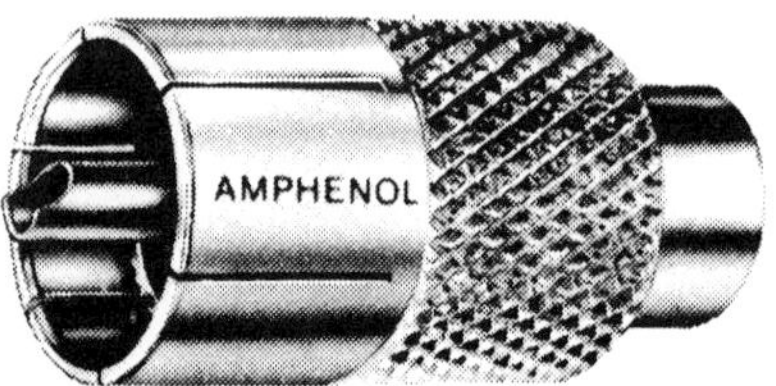

Figure 6–1B Video connector BNC. *(Provided courtesy of Amphenol.)*

All video equipment inputs and outputs are designed to terminate in 75 Ω, their nominal impedance. Most are unbalanced, that is, nonsymmetric, where one side of the line is at ground potential and that side is the braid, foil, or connector sleeve side of the line. The center connector is the electrically high side of the line. Often video components have *looped* inputs, in which the signal comes in on one connector and leaves on an adjacent connector, looped to feed the same signal to another device. If the signal will not be looped to another device, the signal input is terminated in a 75-Ω resistive connector at that point. To avoid mismatches and *standing waves* on the video line, the line should always be terminated in its characteristic impedance (75 Ω) at its farthest point. Standing waves are signal wave reflections, with the reflections bouncing back against each other, on a video line, causing distortion. Standing waves are caused by impedance mismatch.

AUDIO EQUIPMENT INTERCONNECTION

In audio wiring, the signal voltages are carried in shielded cables, that is, in a twisted pair of wires, with the pair covered by a braided copper shield. This is exemplified by mike cable, in which the cable pair carries the signal and the shield is at signal ground potential. Most other wiring in an audio system is unshielded, and it is good practice to keep audio lines separated from lines carrying ac power.

Connectors used in audio work include the XLRs for mikes in balanced circuits, PL 55s of the two- and three-circuit varieties, miniature Pl 55s, and RCA pin plugs and jacks (Fig. 6–2).

Audio amplifier equipment is terminated normally in an impedance of 600 Ω, input and output, excepting only mike outputs and preamplifier inputs, which terminate in 50 to 150 Ω nominal impedance.

Audio and video cables should be looped and coiled in 1-foot circles, and masking taped or tied when the cable is not in use. Video multipin connector cables and camera cables should be looped in figure eight configuration, to avoid tight bends and to facilitate handling, and should be tied at the crossing point of the figure eight during storage.

IMPEDANCE AND IMPEDANCE MATCHING

Impedance is a current flow-resisting characteristic of every ac-operated electronic device. Impedance is denoted by Z, and it is measured in ohms.

For a maximum transfer of signal energy and minimum distortion between electronic components, their impedances should match. That is, the output impedance of the *source*, or feeding component, should be the same as the input impedance of the *load*, or accepting component.

If there is an impedance mismatch between components, in addition to a signal gain loss, there will be a frequency response loss which will degrade the signal quality. If, however, the mismatch is very great (and deliberately so) between a low-Z output and a high-Z input, say 10,000:1 or more, we say that the input is *bridging*. In a bridging situation, there will be a signal gain loss but no frequency response loss. The loss in signal level, then, must be made up by amplification in the bridging device or by an amplifier following.

PATCHING

Patching is a system used to interconnect or disconnect two or more pieces of video equipment or two or more pieces of audio equipment which are perhaps physically adjacent in an equipment rack but are not hard-wired, *normalled through*, to one another. The individual equipment, amplifiers, signal processors, or other devices have their inputs and outputs wired to jacks on a video patch panel or an audio patch panel. Separate short connector cables, with identical plugs at each end, are used to interconnect the devices. On the patch panel there are *mults* as well, or groups of floating jacks connected only to each other (Fig. 6–3). The mults can be interjected between devices, so that an output of one device can feed several inputs of other devices. There are also 600-Ω (audio) or 75-Ω (video) terminating impedances (resistors) which can be

Figure 6–2 Audio connectors. *(Provided courtesy of Switchcraft.)*

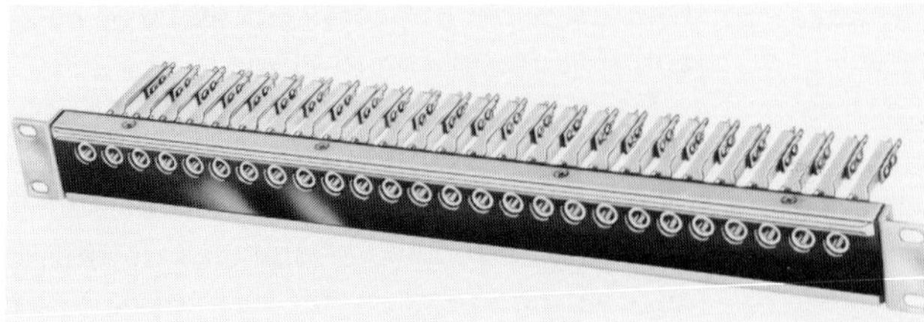

Figure 6–3 Audio Patch Panel *(A)* and Audio Patch Cord *(B)*. *(Provided courtesy of Switchcraft.)*

A

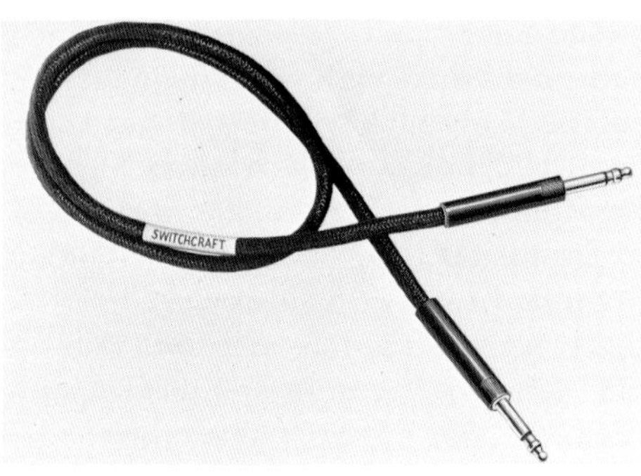

B

patched with equipment components in either series or parallel to make necessary impedance matches. The primary purpose of the patch panel, aside from connecting the devices, is to enable a variety of configurations of equipment without the necessity of rewiring and to maintain impedance matching or bridging between the devices thus connected.

Theory tells us that two same impedances connected in parallel halve the total impedance; and two same impedances connected in series double the total impedance. Thus some simple mathematics will aid in connecting devices so as to avoid an impedance mismatch.

DECIBELS, VOLUME UNITS, AND PEAK-TO-PEAK VOLTS

Reference often is made to *energy levels* in discussion of audio or video equip-

ment. Audio energy levels are commonly stated in volume units (VUs), decibels (db), and in dbm (decibel, 1 milliwatt). Video levels are stated in volts, positive or negative, peak to peak.

By definition, 1 db is 0.1 bel (B), which is named for Alexander Graham Bell. The bel is too large a unit for ordinary usage, so the decibel is used. Decibels may be written in lower case, as we do, or they may be written dB. Either is correct. Decibels are a logarithmic rather than algebraic or arithmetic function. A good rule of thumb to remember regarding decibels is that an increase of 3 db doubles the gain whereas a decrease of 3 db halves the gain.

Decibels and VUs are of relative (not absolute) size. The decibel is an expression of a ratio of two voltages, two currents, or two power values.

A reference level must be stated, in each case, to which the decibel measurement is referred. Thus the decibel itself is referenced to 6 milliwatts (mW) of power in a 500-Ω line, which happens to be an obsolete reference base. The dbm, a more current measurement, is referenced to 1 mW of power in a 600-Ω line, which has a sine wave root-mean-square value of 0.774 V across it.

Both the decibel and the dbm measure sinusoidal power, such as that emanating from an audio tone oscillator. A sine wave has equal positive and negative values above and below a fixed baseline.

One VU is a change of 1 dbm for a *com-*

plex waveform such as speech or music, as differentiated from a sine wave.

BLOCK AND FUNCTIONAL DIAGRAMING

Often we describe electronic circuitry by using the standard engineering methods of block and functional diagraming. Block diagrams use blocks, oblongs, and triangles tilted on one side to picture circuits. The blocks indicate components, the tilted triangles indicate amplifiers, and lines between blocks indicate wiring, with arrowheads on those lines showing the direction of signal flow.

Substituting schematic diagram symbols in some instances for blocks increases the value in understanding the *functional* diagram, as it is called. In both cases, the diagram helps us to understand how the circuitry functions (Fig. 6–4).

AMPLIFIERS

Many fine texts are devoted solely to electronic amplification and amplifiers. We describe them here in a very cursory manner.

An amplifier's function is to increase, or enlarge, a signal passing through by a fixed amount, changing only the gain of the signal, without changing or distorting its other parameters, such as its frequency response.

Amplifiers are the *active* components of electronic circuits. Other components are described as *passive*. Amplifiers can be described by either class or function. An amplifier's class—A, AB, B, or C—describes its electrical relationships, or, more precisely, that portion of its input cycle during which certain currents are permitted to flow under full-load conditions. An amplifier's load is the circuitry into which it feeds, or "works." Amplifiers are designed to increase either signal voltage or signal power.

Describing amplifiers by function indicates what they do. They can be audio preamplifiers, which are voltage amplifiers whose design parameters are such that they have very high gain—in the area of 30 to 50 db—with very low inherent noise and distortion. They can be power amplifiers, which are designed to deliver many watts of power to a load. Power amplifiers are often called *current amplifiers*.

Amplifiers are often *verbally* dissected for discussion by referring to the *stages* of amplification within them.

In discussing how much work an amplifier can do, we discuss at the same time how cleanly it does that work by referring to its *distortion level*. An amplifier's variance from absolute purity of reproduction is stated in terms of a number of factors. *Headroom* is the factor that describes the limit on gain that an amplifier can tolerate before it will clip the tops of a waveform passing through. The *slew factor* measures the amplifier's transient re-

Figure 6–4 Block and schematic diagram symbols.

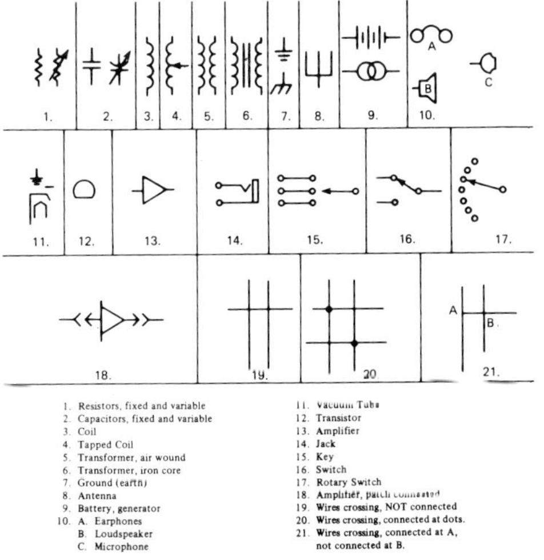

1. Resistors, fixed and variable	11. Vacuum Tube
2. Capacitors, fixed and variable	12. Transistor
3. Coil	13. Amplifier
4. Tapped Coil	14. Jack
5. Transformer, air wound	15. Key
6. Transformer, iron core	16. Switch
7. Ground (earth)	17. Rotary Switch
8. Antenna	18. Amplifier, patch connected
9. Battery, generator	19. Wires crossing, NOT connected
10. A. Earphones	20. Wires crossing, connected at dots.
B. Loudspeaker	21. Wires crossing, connected at A,
C. Microphone	not connected at B.

sponse to sudden intense, steep rises in frequency. The *transient overload recovery time* measures the amplifier's ability to settle back to normal operation after having been overdriven. And *frequency response* is a measurement which includes a description of the portion of the spectrum to be amplified versus the degree of uniformity over that portion. Finally, the *signal-to-noise (S/N) ratio* measures the inherent noise of an amplifier in relation to its signal gain. The greater the S/N ratio, the cleaner the amplified signal.

CABLE BUILDING

The most frequent source and cause of equipment failure in a television system are its cables, most particularly at the connector terminations. And the connectors of the audio subsystems suffer more problems than do the video connectors. It is important, therefore, to know how to build cables, that is, how to place or replace connectors on cables.

The audio cables that concern us most are mike cables. These connectors are termed *XLRs*, and they are almost a standard in the industry. They were introduced by the Cannon company, years ago, to replace the much larger P type, still found in ancient audio installations. The XLR-type connector is now made also by Switchcraft in their A3 series and by Neutrik of Switzerland.

Similar to other connector pairs that "mate" to complete a circuit, XLRs come in male and female configurations. It *is* standard in the business for mikes to terminate in a male connector. So if the mike has a male connector, then its cable must have a female connector at the mike end and a male connector at the other end, to mate with either another (extension) ca-

ble or a wall receptacle or equipment input connector. Wall receptacles and equipment *input* connectors always use a female connector, whereas equipment *outputs* always use a male connector.

The XLR connector which is used for audio work, whether male or female, has either three or four solder connections at its back end, its front end being the "mating" end. Other XLR connectors having a greater number of terminations are used for different circuitry connecting purposes. In the audio XLR, the terminations, or *pins*, are labeled pin 1, pin 2, pin 3, and shell, the connector-cover metal enclosure. Pin 1 is always ground (earth); sometimes it is connected also to the shell and to the cable-shielding braid. Standardization falls apart here, but in most cases pin 3 is the *low* electrical side of the circuit and pin 2 the *high* side.

The importance of this information is twofold. First, when plugging a mike into an equipment-connected cable results in a distortive "hum" in the system instead of audio, it is an instant indication that there is an open-circuited (broken, disconnected) ground. The wire to pin 1 in the cable, at one of its ends, has disconnected from the pin. It is possible that the open circuit is in the mike or receptacle, but infinitely more likely to be in the cable because of the hard physical usage that cables get.

Second, if a cable's XLR connector were wired with pins 2 and 3 reversed in polarity (the wrong wire to the right pin), then the mike connected to that cable and connector would be electrically *out of phase* with any other mikes connected to the same audio mixer. The misphasing of two mikes connected to the same audio mixer would cause the mikes to electrically subtract rather than add, in the amount of audio signal contributed to the system.

Heavy-duty mike cables manufactured long ago have three internal flexible stranded copper wires, each covered with rubber and usually color-coded white, black, and brown. They are surrounded by a braided copper shield which protects against intrusion of stray external electric fields. The shield is covered by a rugged black or brown neoprene or rubber jacket. The newer mike cables have but two inner wires, usually color-coded red and black, surrounded by an aluminum foil shield which also acts as the ground wire, and jacketed in one of many colors of neoprene or fabric.

There is no industry standardization of wire color coding to XLR pins. But it is most important to remember that within a given audio system or installation, standardization must be maintained to avoid inadvertent phase reversals.

The general rule observed by this author is that in a three-wire cable, the white wire should be ground (pin 1), and the other two wires as appropriate to the system. What is appropriate to the system may be easily checked by releasing and sliding back the shells of *two other* mike cable's male ends (males are easier) enough to expose to view the wiring color code in use. The cable shield should be connected to the ground pin and also to the shell connector (if there is one).

In a two-wire cable, color-coded red and black, red should be connected to pin 2, black to pin 3, and the shield to pin 1 and the shell connector.

A primary reason for problems occuring at the connectors of mike cables with hard use is that, despite a strain relief at the back of the connector, often there is too much individual wire slack within the connector shell. The individual wire lengths coming from the body of the cable to the individual pins should never exceed 0.5 inches or should be only long

enough to reach the inside of the connector pin.

In most cases, video cable connectors are the simplest to construct. Single video cables have a center conductor which *is* the connector center conductor or is soldered or crimped to the connector center conductor. The shield is soldered or banded and crimped to the connector shell. It begins to get complex when a camera cable or similar multipin connector is constructed. On multipin connectors, the pins are always numbered, and they are connected to specifically color-coded wires. Most often the coaxial cables are in the very center of the connector, and the coaxial cables are connected first. The rest of the wires are connected in a circular fashion from the center out to the perimeter of the connector. Neatness counts. Where wires are soldered to pins that are very close together, extreme care should be taken not to short-circuit wires or pins with excess solder or frayed wire strands.

TROUBLE DIAGNOSTICS

Here we describe a simple logical system that explains where to look, and why, when equipment appears not to be working. This is very different from workbench repair of equipment, the subject for yet another text. This diagnostic system has saved the day for many equipment users over a long period. Indeed, when a major equipment component has failed, this system can be used to pinpoint the failure for the maintanance technician, thus saving precious time.

Equipment components (camera, VTR, audio mixer) should always be set up and checked out *by themselves* before they are connected by cable to other components. That is basic to the logical system.

When an entire chain or group of components is inoperative, it eliminates the question of Which one? or What cable?

Here are some examples:

A video camera is "fired up" and appears not to be working. First, go to the simplest possible cause, before you assume the worst. Check that the lens cap was removed and that the filter disk is not turned to the CAP position. Next check the power source. Is there a low-battery indicator in the viewfinder? What does it tell you? If the camera is connected to a source other than battery, was it turned on and have the circuit breakers been checked? Finally, what visible indication is there that a camera is or is not working? Check the viewfinder controls (brightness, contrast). Then look for the picture, or turn-on bars. It is always the simplest problems that cause the most trouble.

Here is an audio example: Five mikes on a stage or at a congressional hearing are to be fed to an audio mixer. We always assume from the outset that there may be trouble with the mikes, their cables, or the mixer. This is how we logically connect them, so that we can diagnose any problems as we proceed. The mixer is placed on a table and connected to a power source. Earphones are plugged into the mixer. The mikes are each placed at their pickup points and are connected to cables. The cables are run to the mixer, marked, but *not* connected. Now, we connect the system and check it out at the same time. We put the earphones on, power up the mixer, and turn on the internal tone oscillator. Hearing a tone indicates that the mixer is operative. No tone indicates trouble in the mixer or earphones. We look at the VU meter. Is it illuminated? Does it indicate tone? If it does and no tone is heard, we try another headset. Next we connect each mike ca-ble in turn, open the mike pot, and listen. If nothing is heard, we replace first the cable and then the mike. We close the pot and go on to the next mike. When all the mikes and their cables are connected and checked, the system is verified as ready to use.

The third example involves the electronic news gathering (ENG) connection of a camera and a VCR. The camera is checked out as in the first example. Then the VCR is checked independently. A monitor is connected, and the power is turned on. A modulation source other than the camera is fed from the VCR to the monitor. This could be a recorded tape, which verifies the VCR, the monitor cable, and the monitor. Next the multipin camera cable is connected from the camera to the VCR. Bars are fed from the camera to the VCR. If they are received on the monitor, the system is verified operational; if not, a BNC coaxial connection is made from "camera out" to "VCR video in." Reception of bars with this single cable indicates a bad multipin cable.

These examples describe the logical diagnostic system. First, verify one central component. Then expand it one segment at a time to include all peripheral components and connecting cables, until the system is complete. There could be many more examples, some even more illustrative. Perhaps technically minded readers will refer their personal examples to the author through the publisher, and these will find their way into subsequent editions of this text.

REVIEW QUESTIONS

1. Describe coaxial cable.
2. What is a balanced line?
3. Explain impedance. How do we denote it?

4. Describe source and load. What is bridging?

5. Why do we patch?

6. What is a dbm? A VU? How do they differ?

7. What does an amplifier do?

8. Describe headroom, frequency response, and the *S/N* ratio.

9. What is an XLR? Why do connectors mate?

10. What is color coding?

11. Describe the benefits of a logical diagnostic system.

Chapter 7

The Video Tape Recorder

A BRIEF HISTORY OF VTRS, TYPES OF VTRS USED IN TELEVISION, MODES OF LOADING

In the 1940s, when a *copy* (duplication) of a live television program was needed, either for archival purposes or for a network repeat to a different time zone, the *live* program was fed online to a film chain camera which made a *telecine* copy of the program on movie film. This process was expensive, had relatively poor film quality, and required the time delay of film development and printing, which in turn determined how soon the copy could be aired.

Audio tape was already in use, having been developed right after World War II. And in 1950 the facilities and funding of Bing Crosby Productions were used to modify an audio tape recorder so that a 1-inch-wide tape running at a speed of 100 inches per second (ips) produced a crude video picture on an operable video tape recorder (VTR).

RCA's engineers then built a ½-inch-tape VTR which ran at a speed of 360 ips. At these very high tape speeds, compared to 7.5 or 15 ips for audio recording, an immense amount of tape was devoured in a very short time. And the research en-

gineers found that the high relative tape-to-record head speed was necessary to magnetically print the high video frequencies on tape.

On both the Crosby and RCA video tape recorders, the magnetic imprint was written in a longitudinal fashion, called *longitudinal video recording* (LVR), just as in audio recording. At the time it was quite obvious that tape quantity was going to be a limiting factor in the development of video tape recorders.

Next the engineers at the Ampex Corporation, a major manufacturer of audio tape recorders, designed a VTR with four recording heads *mounted on a rotating cylinder*, which wrote the magnetic information on the tape *diagonally*. With the head cylinder rotating at 240 revolutions per second and 2-inch-wide tape running at 15 ips, an effective magnetic writing speed of 1800 ips was reached.

In the Ampex system, each frame of picture information was divided among the four heads, and the system was called *quadruplex*, or "quad", recording. It was first designed for black-and-white video recording and then expanded to record color in 1959.

In the late 1960s, the Sony Corporation marketed the two-head rotating-cylinder helical-scan VTR that used half-inch-wide tape for its black-and-white picture, small recorder called the Portapak. This development brought the mobility needed if the VTR was going to be used out on the streets for news gathering. Helical-scan recording was improved and standardized in 1969 by the Electronic Industries Association of Japan (EIAJ) and was called *½-inch EIAJ*.

In 1972 the Sony Corporation simplified and added increased frequency response and stability to helical-scan recording by introducing the ¾-inch U-Matic format video cassette recorder.

However, although helical-scan recorders were lightweight and thus portable, their tapes were not compatible with the NTSC system quadruplex machines in use at U.S. television stations and were not stable enough for broadcast use.

It took the design of the *time-base corrector* (TBC), developed in 1972, to stabilize the helical-scan picture. Then helical-scan recorders had the precise horizontal line timing that allowed them to conform to NTSC standards and permitted helical-scan recorded tapes to be used interchangeably with studio quadruplex machines, re-recorded on the quadruplex machines. This opened the use of helical-scan recorders to broadcast stations and networks which had previously made a tremendous capital investment in quadruplex equipment.

In 1975 the Sony Corporation introduced a new television tape format, to replace open-reel tape. It was a self-contained cassette that Sony called Beta, and it employs ½-inch tape. This was followed in 1977 by other companies using a similar format called VHS. Originally both of these tape formats were intended primarily for the then-infant consumer VTR market. The home VTR market has proliferated; and because of their large

supply, easy availability, and low prices, its videotapes are now being integrated into the design of cable system equipment.

Video tape recorders that are used in cable TV are currently of the ¾-inch U-Matic cassette format. Industrial television uses the NTSC or EBU type C high-band helical-scan recorders with 1-inch tape. Commercial broadcasting uses both the type C machines and the quadruplex recorders (no longer manufactured) with 2-inch tape.

U-Matic machines are found in three styles. The first and most lightweight is used for field recording, often lacks sophisticated features, often is a record-only machine, and generally is operated from a battery power supply or from a lightweight ac power supply. Often the field recorder can only record on a small cassette for up to 20 minutes. This cassette, however, may be used in the larger studio machines, by aligning the spine of the small cassette with a special groove in the opening of the studio machine.

The second style of VTR is a record-playback machine that can be free-standing on a desk surface and is ac-powered. The third type of VTR is similar to the second type, but has additional features that are used for tape editing. We discuss tape editing in detail in Chapter 9.

Two types of cassette tape loading are used in VTRs. [Sometimes they are called video cassette recorders (VCRs) rather than video tape recorders.] The top-loader type requires depressing a control button, usually labeled EJECT; when the recorder power is turned on, this causes a loading mechanism to rise from the top of the machine. Then the cassette is placed in or removed from that mechanism. The front-loader type merely requires that the tape cassette be carefully inserted in a slot on the front panel of the machine. An eject button releases the cassette for removal. Care should be taken to insert the cassette straight in, and not at an angle. The top side of the cassette should be facing up, and any labels should be firmly attached, to prevent jamming the cassette mechanism.

There are a number of different uses for the VTR in television. In playback mode, the VTR can be a video source into a video switcher for either program inserts or an entire package program. It can be used as a dubbing source, feeding another recorder to make tape copies; or it can be the insert source in a tape editing facility.

In the record mode, the VTR can record from a single camera or be the terminal point for the output of a switcher. Together with another video tape recorder, an editing controller, and a time-base corrector, the VTR can be part of a video editing system.

As before, we now examine some of the brand-name VTRs, starting with the U-Matic cassette types and progressing to open-reel recorders.

PANASONIC NV-9240 U-MATIC RECORDER

The NV-9240 is the recorder unit of the Panasonic Series 9000 ¾-inch video cassette system, which also includes the model NV-9600 editing recorder and the model NV-A960 editing controller (Fig. 7–1).

The NV-9240 is a helical-scan recorder employing two rotary heads. It has two audio tracks, uses ¾-inch tape cassettes, and has a tape speed of 3.75 ips. The record/playback time is 60 minutes with a NV-P26 tape. The fast forward/rewind time is less than 4.5 minutes for a 60-minute tape. Video horizontal resolution is better than 260 lines for color. The ma-

Figure 7–1A The Panasonic NV-9240 video tape recorder. *(Provided courtesy of Panasonic.)*

chine weights 77 pounds and can be used as a master studio recorder, a dubbing deck, or a source player in an editing system. The NV-9240 uses a direct-drive video head cylinder and direct-drive capstan servomotor for high stability. Its microprocessor permits soft-touch, non-locking pushbutton controls and allows direction mode changes without having to pass through the stop mode. Audio track 1 accepts time code information for use in automatic editing systems. (Time code is discussed in detail in Chapter 9.)

The machine has dubbing input and output connectors, video head switching that is performed during vertical interval, and a separate control-track head that reads tape position for display on the LED tape counter during fast-forward and rewind modes.

PANASONIC NV-9600 EDITING U-MATIC RECORDER

The Panasonic NV-9600 has all the features of the NV-9240 plus two "flying erase heads," which are rotary rather than stationary (Fig. 7–2). It performs frame-by-frame edits. This deck contains a frame

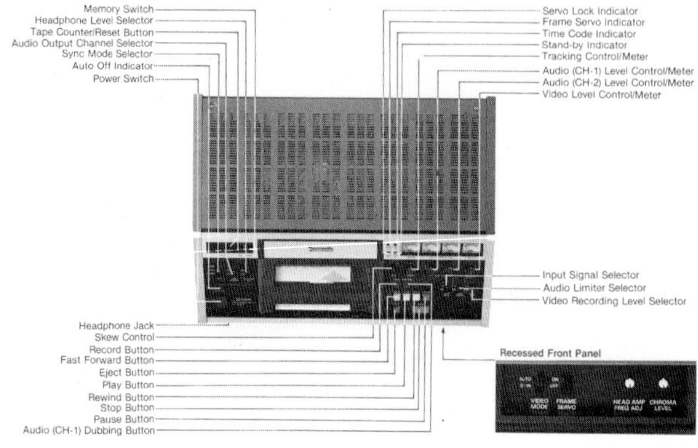

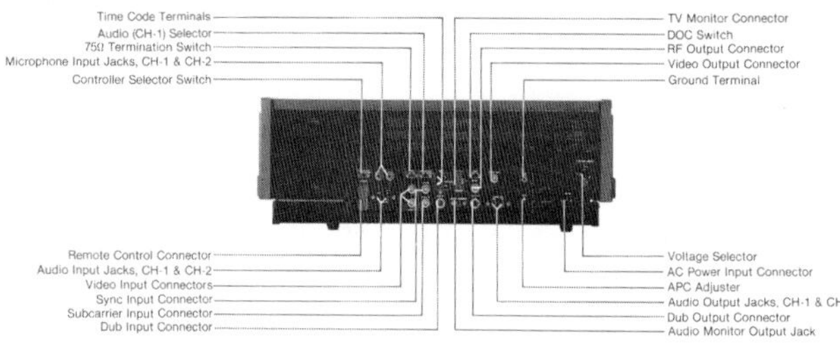

Figure 7–1B Controls description diagrams of the Panasonic NV-9240 tape recorder. *(Provided courtesy of Panasonic.)*

servomotor and a frame indicator which lights when both the editor recorder and the source player fields are in the correct position. Horizontal timing is adjustable to minimize *flagging* at the editing point.

Figure 7–2 Panasonic NV-9600 editing recorder. *(Provided courtesy of Panasonic.)*

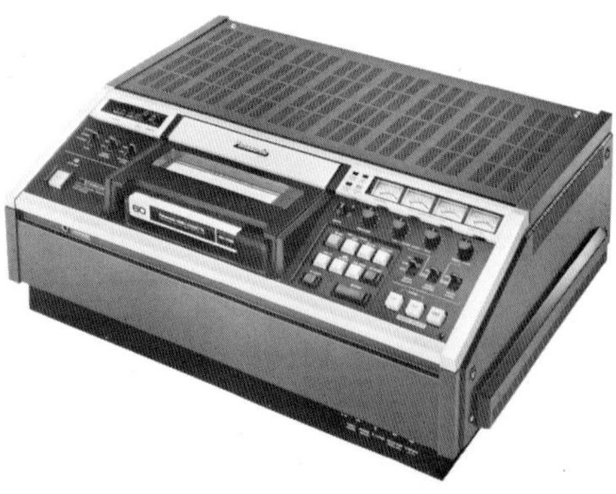

SONY VO-5800 U-MATIC VIDEO CASSETTE RECORDER

The VO-5800 is part of a Sony system that also includes the VO-5850 editing video cassette recorder and the RM 440 editing controller (Fig. 7–3). The VO-5800 is a rotary two-head helical-scan record/playback device. It has a horizontal resolution of 240 lines in color and an *S/N* ratio of better than 46 db. Its tape speed is 3.75 ips, and it has a maximum record/playback time of 60 minutes (with a KCA 60 tape cassette). Its fast-forward and rewind time, end to end, is 4 minutes. It uses U-Matic video cassettes of the KCA or KCS tape types.

The upper half of the front panel has a front-loading cassette compartment, with a skew control on the lower right. The skew control adjusts the tension on the tape. Sliding the control to the right or

Figure 7–3 The Sony VO-5800. *(Provided courtesy of Sony.)*

left eliminates hooking or bending at the top of the picture as a result of improper tape tension.

To the right is the power on/off switch, and below that is the timer switch. Timer-activated recording or playback is begun when this switch is set to record (REC) or PLAY. When it is set to REC, all function buttons except STOP are deactivated.

To the extreme left of the cassette compartment is the audio level control section, comprising a meter and a level control for each of the two audio channels (tracks) on the tape. Below the controls is a limiter switch which is used when audio containing high-level peaks is recorded. For normal audio recording, the limiter switch is left in the OFF position.

Between the audio control section and the cassette compartment is the tracking control, which is used to minimize the tracking variance between two machines. Normally this control is set at the top, fixed position.

On the lower half of the panel, left to right, we see the audio monitor section, a headphones level control and jack, and an audio monitor switch which feeds audio channel 1, audio channel 2, or a mix of both to the earphones. Below the audio monitor switch are two mike input jacks, and to their right two video input selec-

tors. When the second switch is set to LINE, the first input selector chooses between the video 1 connector and the video 2 connector on the rear of the recorder. This second input selector may also choose TV, to record from the TV connector or from dub, to record from the DUB IN connector.

These function buttons follow: the EJECT button; the record indicator and record (REC) button, which must be pressed at the same time as the PLAY button, to record; the DUB/CH 1 audio dubbing button and indicator. Press DUB/CH 1 at the same time as the PLAY button to add sound to a previously recorded videotape on audio channel no. 1. Below these buttons are the STOP, the rewind (REW), PLAY, fast-forward (F FWD), and PAUSE buttons. All but the STOP button have associated indicator lights.

To the right of the function buttons are three indicator lights. (1) The standby light comes on while tape is being threaded or unthreaded from the tape path. (2) The AUTO OFF light indicates, when lighted, that there is moisture condensation in the machine. When this light is on, the machine will attempt to fan-dry itself and will not operate until it is dry. (3) The servo indicator light, to indicate that the tape drum motor and the capstan motor are locked together and that the tape transport mechanism is stabilized.

Above and to the right of these indicators are the programmed operation switch, the search switch and dial, and the memory counter section, all of which are used in videotape editing and whose functions are described in Chapter 9.

On the rear panel of the VO-5800, left to right, we see the power cord, an unswitched AC OUT connector, a ground connector, and a 33-pin female remote connector. This connector accepts either a 33- or a 20-pin male plug to connect a remote-control unit, an RM 440 editing controller, or an RX 353 or 303 automatic search control. Next follows an RF OUT F type of connector, and above it is an RF modulator compartment. With these, a picture can be fed to a conventional TV receiver and seen on either VHF channel 3 or 4. Below these is an 8-pin female TV connector which feeds both video and audio signals to a color monitor.

Next follows the video in 1 and 2 connectors, BNC type, and below them is the seven-pin male DUB IN connector. Below them are seen the video out BNC connector and the seven-pin female DUB OUT connector.

A very important point should be made about the dubbing connectors and their cables now. *Never* connect DUB OUT on one machine to DUB IN on another *at the same time as* the DUB OUT connector on the second machine is connected to the DUB IN connector on the first machine. This will instantly produce escalating video feedback at extremely high levels and will damage the VTRs.

To the right of the video in and out connectors is the subcarrier in connector (SC IN), BNC type, to connect the subcarrier from a time-base corrector. Also there is the TBC switch, which is set ON when a time-base corrector is connected.

The final set of connectors are the audio line in jacks for two audio channels, the audio monitor jack, the two audio line out jacks, and the RX data jack for recording and reading the data recorded on the tape by the RX 353 automatic search control.

SONY VO-5850 U-MATIC VIDEO CASSETTE RECORDER

As in the less sophisticated model (the VO-5800), on the Sony VO-5850 the lower front panel swings out for ease of opera-

tion. The VO-5850, too, is a front loading U-Matic system, which can be rack-mounted or free-standing on an editing desk (Fig. 7–4).

The controls of the 5850 can be best described by their functional grouping: The cassette compartment is located on the upper right side of the panel. To its right is the power off/on switch. On the lower right side of the cassette compartment is the skew control, a slide switch which adjusts tape tension. It may be moved to the left or right to get a normal picture *during playback only*, when picture "hooking" distortion appears in the upper part of the viewing screen. Normally the skew control is set to the center detent position, and it returns automatically to this position when the VO-5850 is in the record or edit mode.

To the left of the cassette compartment are the audio level and the video level and tracking sections. There are two audio level meters, one for each of two audio tracks, which indicate level during either recording or playback. Below are two separate audio channel level controls and, below them, an audio limiter switch. The limiter switch should be set to ON only for audio input with high-level peaks. For normal-level audio recording, the switch is set to OFF.

Figure 7–4 The Sony VO-5850 VCR front view. *(Provided courtesy of Sony.)*

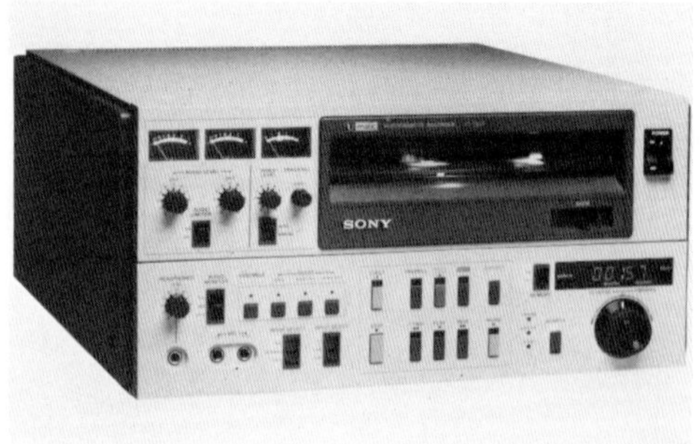

To the right of the audio control section on the upper panel are the video and tracking controls and meter. This meter indicates the video level during recording and whether tracking is correct during playback. Below the meter are the video level control and the tracking control. With the automatic/manual switch set to MANU, adjust the control so that the video recording level is within the blue zone on the meter. In the AUTO position, the video recording level will be set automatically.

The tracking control is adjusted to minimize tracking variance between machines, and so the tracking meter deflects as far to the right as possible.

The lower part of the panel, from left to right, shows the audio monitor section comprising the audio monitor switch, the headphones level control, and the stereo headphones jack. The switch selects sound through the headphones, or monitor speaker: channel 1 only, channel 2 only, or the center position MIX, which puts channel 1 on one earphone and channel 2 on the other, or mixes the two channels to the monitor speaker. Below the audio monitor switch are two mike input jacks, one to each audio channel.

To the right are the editing mode selectors. First is the assemble button and indicator, followed by the three insert buttons and indicators. Two or more insert buttons may be pressed at the same time. The functions of these buttons are described in Chapter 9. To cancel a mode selected by one of these buttons, press the button a second time. The indicator will go out. To change editing modes, first cancel the selected mode and then reselect.

Below the editing mode selectors are the sync mode and input selector switches. The sync mode switch selects operation with a time-base corrector on the output,

normal recording or playback, or operation in the editing mode. The input select switch provides recording from the TV connector, line recording from the video in connector, and recording from the DUB IN connector. All these mentioned connectors are on the back panel of the VO-5850.

Next, to the right, are the tape machine's function buttons in two rows: The top row has the EJECT button; the PREROLL button and indicator; record (REC) button and indicator, which must be pressed at the same time as the PLAY button; EDIT button and indicator, which must be pressed simultaneously with PLAY to enter the editing mode, and the CUT OUT, button which is used to end an edit or cancel the record mode. The bottom row has the stop (STOP), fast-forward (F FWD), PLAY, rewind (REW), and PAUSE buttons.

To the right are three indicator lights: (1) The standby light comes on while tape is being threaded or unthreaded on the tape path. (2) The AUTO OFF light indicates that there is moisture condensation within the recorder. When it is lighted, the recorder will not operate. (3) The servo lights to indicate that the tape transport is stabilized and the tape may begin to move.

The final operating section is the search control section and the memory counter. Both are used extensively and described in Chapter 9.

The rear panel of the VO-5800 has all the connectors that attach this equipment to its inputs and outputs: From left to right: An AC IN connector accepts the ac power cord, and an AC OUT connector enables powering adjoining equipment from the same power source. A ground terminal reduces hum in the audio signal. The 33-pin remote connector permits connection of an optional editing con-

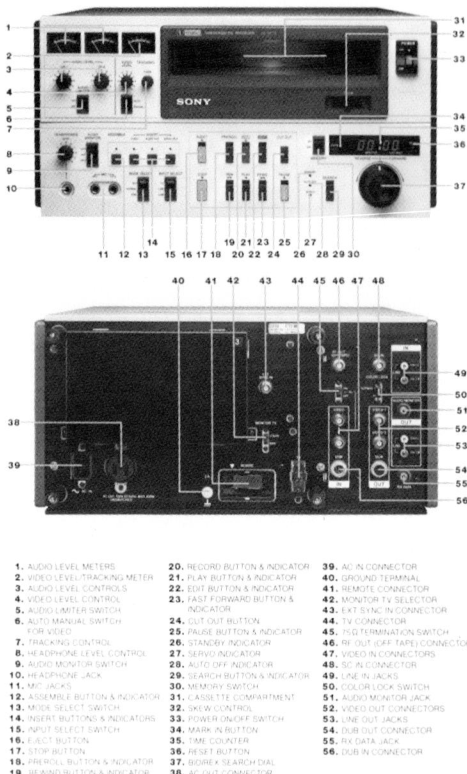

Figure 7–5 The Sony VO 5850 controls description. *(Provided courtesy of Sony.)*

troller, an automatic search controller, or a remote-control unit for the VTR. This is a female receptacle, which means that the connecting plug must be male. A 20-pin remote connector, male, also may be mated with this connector without an adapter.

Above this is a monitor TV switch which selects either black-and-white or color monitor. Above this switch is the EXT SYNC IN connector, a coaxial cable (BNC type), which connects the signal of a time-base corrector or an external sync generator to the VTR. The recorder will synchronize with this incoming signal.

Next there is an eight-pin female TV connector, which connects the VTR to a video monitor. Video input and audio input to the monitor are made with this connector.

To the right of the TV connector, top to bottom, we see the RF OUT (OFF TAPE) connector, BNC type, which supplies the tape output to a time-base corrector; the 75-Ω termination switch, normally set to ON, except when the *loop-through* output of the video in connector is used, in which case it should be set to OFF.

Below the termination switch are the two looped-through video in connectors, BNC type, one of which carries the video signal to be recorded into the VTR. Below the video in connectors is the DUB IN (male) connector for feeding input from another VTR *for direct copying* from machine to machine.

To the right, top to bottom, are found the subcarrier input (SC IN) connector, a BNC which connects the subcarrier from the time-base corrector to the VTR; the color lock switch, set to NORMAL unless the playback picture has no color or has an abnormal hue, when it is set to the position marked with a dot.

Below are two video out BNC connectors, in a looped-through configuration, and below them the DUB OUT seven-pin female connector to feed input to another VTR for direct copying.

The last column of connectors, from top to to bottom, is as follows: The audio line in jacks, phonograph type, feed channel 1 and channel 2 audio in. The audio monitor jack, mini type, feeds the audio signal selected on the audio monitor switch on the front panel to an audio monitor. The audio line out jacks, phonograph type, are the audio output from recorded tape. Finally, the RX data jack, mini type, is for recording, and reading the data recorded on tape, by an RX 353 or RX 303 automatic search control.

Let us repeat a caution stated for the

VO-5800. When you use connecting cables with the VO-5850 VTR, the DUB IN and DUB OUT connectors of two VTRs should *never* be connected in parallel (that is, one machine's output to another machine's input and *at the same time*, the first machine's input to the second machine's output).

To Operate the VO-5850

First connect the VO-5850 by the correct cables to a video monitor and to whatever inputs or outputs will be necessary. Then turn on the video tape recorder's monitor power.

Turn on the VTR power, and insert a recorded video cassette.

Set the monitor input switch to VTR or LINE, depending on what cable is used to feed the monitor. Adjust the monitor for optimum picture, and set the VTR monitor switch to either color or black-and-white.

To Record

Set the mode select switch to NORMAL; then the input select switch to TV, line, or dub, whatever the requirement. If it is TV, select the channel; if a camera line input, adjust the camera; and if a dub, feed the tape to be duplicated.

Adjust the video recording level, by either setting the automatic/manual switch to AUTO or by setting it to MANU and operating the control knob to keep the meter needle in the blue zone on the scale.

Adjust the audio recording level by operating the audio level knobs, so that the meter's needles peak at zero, with the audio limiter switch off.

Set the timer to 00 00 with its reset button.

Insert a clean, unrecorded tape cassette in the machine.

Press the REC and PLAY buttons simultaneously to start recording. When recording is finished, press the STOP button.

If recording continues to the end of the tape, the tape will automatically rewind to the beginning and stop.

To Play Back

With all peripheral equipment correctly attached, turn on power and insert a recorded video cassette.

Set the mode select switch to NORMAL or to TBC if a time-base corrector is attached.

Set the audio monitor to either mix or the desired audio channel. Note that this monitor setting has no effect on the audio output from the line out jacks.

Press PLAY to begin playback and STOP to stop playback.

When the tape end is reached, it automatically rewinds to the beginning and stops.

TIME-BASE CORRECTORS AND EXTERNAL SYNC SYSTEMS

It is often desirable to use a time-base corrector (TBC) on the output of a helical-scan VTR, for maximum sync and control-track stability, or the best possible picture (Fig. 7–6). The outputs of the VTR that feed the TBC are the video and RF OFF TAPE, and the outputs of the TBC that feed the VTR are the subcarrier output and the ADV (advance) SYNC TO EXT (external) SYNC IN.

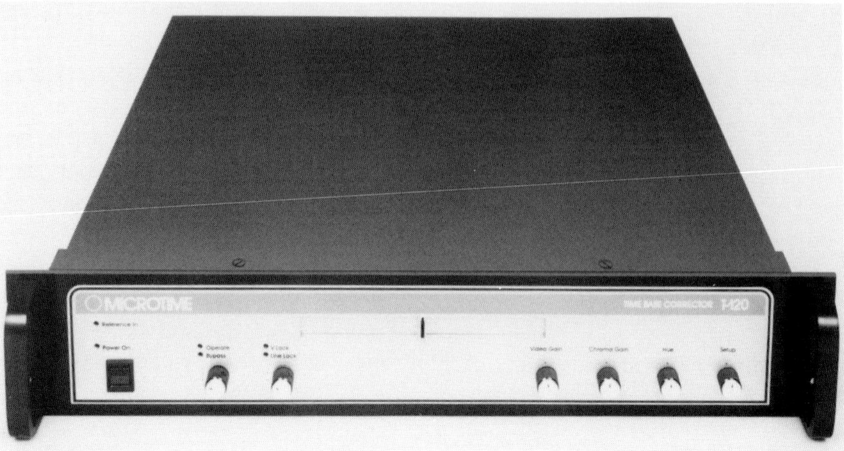

Figure 7–6 Time Base Corrector. *(Provided courtesy of Microtime.)*

An external sync signal can be used, depending on the setting of the mode select switch and the input signals: If the external signals reaching the VTR are a sync signal and a video signal and the VTR is in record, then the VTR will use the sync of the incoming video signal. If the VTR is in playback, it will use the setting of the mode select switch: TBC (sync with sync signal), or EDIT (sync with video signal).

If the external signal is video signal only, the VTR will sync with the incoming video.

If the incoming signal is sync only, the VTR will sync with the sync signal. If there is no input signal, the VTR will sync with the internally generated sync signal. See Fig. 7–7.

NTSC C-TYPE RECORDERS

The Ampex Corporation has joined with Nagra, the Swiss makers of premier portable audio tape recorders, to design and produce the unique Ampex Nagra VPR-5 videotape recorder/reproducer (Fig. 7–8). The VPR-5 is a helical-scan type C portable recorder weighing less than 15 pounds, including 20-minute tape reels, battery, and cover. This low weight is im-

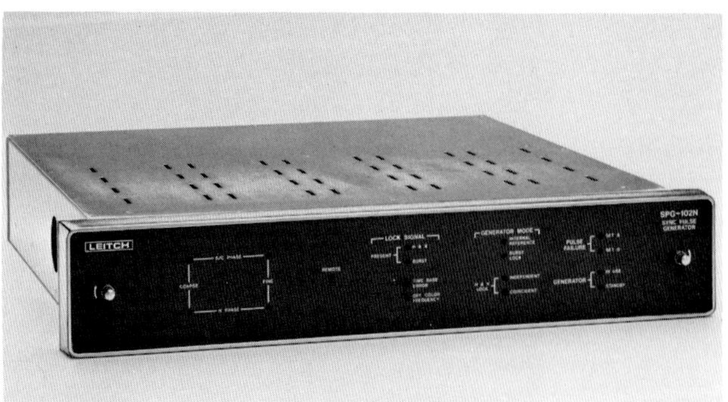

Figure 7–7 A sync Generator. *(Provided courtesy of Leitch.)*

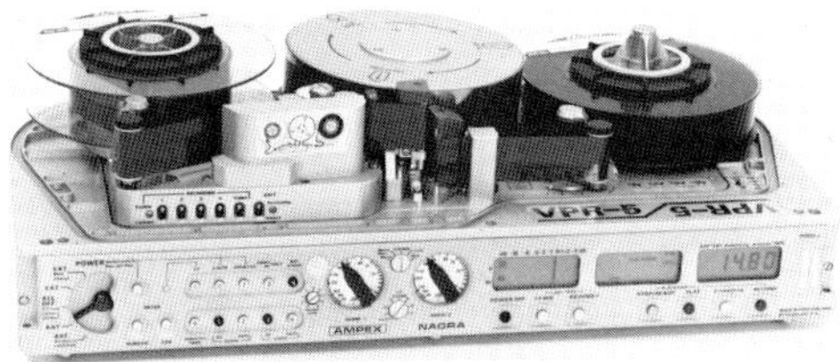

Figure 7–8 Ampex Nagra VPR-5 *(Provided courtesy of Ampex.)*

portant to portable VTR operation when most other portable recorders weigh three to four times more than that.

Because of its Nagra connection, the VPR-5 has very advanced audio technology. There are three audio inputs, with the third being either a line input or a Society of Motion Picture and Television Engineers (SMPTE) or European Broadcast Union (EBU) time code generator input. Metering for all modes of the recorder is by liquid-crystal display (LCD), and for either PPM (peak program meter) or VU audio mode switchable metering.

A variety of interfaces are available for different microphone types and for line inputs. There are seven switch-selectable audio filters to match scene acoustics and to provide low-frequency reverberation cancellation.

For video editing in the field, simultaneously depressing the forward (FWD) and rewind (REW) buttons causes the machine to recue to the beginning of the last take, for review and possible reshooting.

The selected editing mode of either assemble or insert editing is activated by depressing both PLAY and REC together. To review a scene, with automatic recue to the *end* of the edit, STOP and PLAY are depressed together.

Video signal inputs and outputs are by BNC connector at 75-Ω nominal impedance. The recording time is 20 minutes with special 5.5-inch reels at a tape

speed of 9.6 ips. With the cover removed, the VPR-5 can use 9-inch reels for 1-hour recording; note, however, that the battery capacity is 1 hour. There is an optional ac adapter, and the recorder can be rack-mounted for van use or wall- or table-mounted. The video writing speed is 1009 ips.

The Ampex VPR-20B is a portable, reel-to-reel, 1-inch helical-scan machine (Fig. 7–9). It is a type C recorder; as such, its tapes are interchangeable with any other type C machine.

The first thing one notices about most portable C machines is that they appear to have only one tape reel. Actually there are two reels, one atop the other, with the feed (supply) reel on the bottom and the takeup reel on top. To the left and below the tape reels is the tape head assembly; directly below the reels is the operating controls panel.

The tape is threaded in its path by hand

Figure 7–9 Ampex VPR-20B video tape recorder. *(Provided courtesy of Ampex.)*

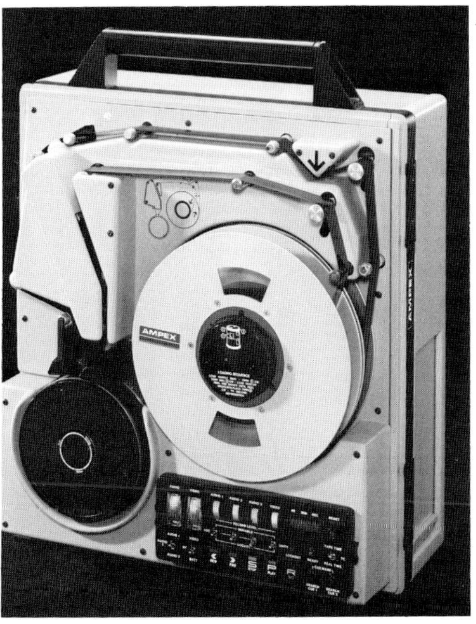

from the feed reel to the takeup, following a diagram printed on the front panel.

There is a cover for the front face of the machine which protects it and its tape in adverse weather conditions. The operating controls are located behind a sliding transparent door, which can be closed to prevent dust or moisture from entering the machine during operation.

These are some of the features of the Ampex VPR-20B: The operating control panel can be used in remote mode for control room use. The camera interface permits tape-remaining and low-battery displays to be shown in the camera viewfinder. When in the record mode, the machine automatically backspaces three frames, for an assemble edit. In the record mode, the viewfinder is fed from a playback head, *off of the tape,* so that what the camera operator sees in the viewfinder is what is on the tape. Further, an optional color stabilizer module is available which permits color video to be seen in the viewfinder or other color monitors. Also optional is a SMPTE/EBU time code generator module, which permits time code to be recorded on audio channel 3. There is a switchable recue to the beginning or end of the last recorded segment, which employs a built-in electronic tape timer. Finally there is three-way powering: a plug-in battery, an external dc input, and an ac power pack that plugs into the battery compartment.

There are three audio channels, with an *S/N* ratio of -56 db on channels 1 and 2, and of -50 db on channel 3, which is normally used only for time code.

The VPR-20B weighs 53 pounds, which includes the battery pack and a 9-inch, 60-minute reel of tape. Its rewind time is 4 minutes for 1 hour of tape. The battery capacity is up to 70 minutes. The tape speed is 9.6 ips, and the video writing speed is 1009 ips. (See Fig. 7–10.)

Type C helical-scan machines are manufactured in studio console or nonportable configurations as well, with the major appearance difference being that the reels are separate, as in audio tape recorders. Studio consoles weigh more than the portables, ranging from about 150 to 300 pounds.

The VPR-3 is a console-mounted, highband, type C helical-scan recorder/reproducer (Fig. 7–11). It has an overbridge which contains a small color monitor and both a vectorscope and waveform monitor. The VPR-3 features a pinch-rollerless design, employing a vacuum capstan to control tape contact instead. Its operating controls panel has both audio and video metering as well as a large-dot addressable fluorescent readout display with direct access. Six menu-identified keys can be used to organize, display, and simplify operational control and the machine's internal diagnostic functions.

The VPR-3 is computer-controlled for many functions, including optimization of video and audio record parameters and five-machine editing. The control of the tape is so accurate that the tape position can be specified in TV lines, instead of fields, for extreme tape-parking accuracy.

Ampex AVR-3 Quadruplex Recorder

The Ampex AVR-3 is a 2-inch quadruplex recorder which no longer appears in the Ampex catalog (Fig. 7–12). Quadruplex machines are obsolete technology but are still found at broadcast stations.

VIDEOTAPE

Videotape, like its relatives, audio tape and computer tape, consists of a plastic polyester or Mylar ribbon that is coated

Figure 7–10 Ampex VPR-20B Anatomy photo. *(Provided courtesy of Ampex.)*

Figure 7–11 The Ampex VPR-3 Console.
(Provided courtesy of Ampex.)

Figure 7–12 The Ampex AVR-3 Quadruplex
Recorder/Reproducer. *(Provided courtesy of
Ampex.)*

on one side with a metallic oxide. The
oxide has the capability of being magne-
tized. That is, a magnetic pattern may be
impressed into its molecular structure if
the oxide is brought in contact with a
magnetizing device such as a video recording head. Similarly, the tape may describe or write that magnetic pattern if it
is brought in contact with a video playback head.

Of various oxide emulsions used to coat
the ribbon base, some are more capable
of magnetization than others. Some have
specific properties that others lack, such
as broad frequency response, but all emulsions have tradeoffs; for example, some

have high abrasiveness, which shortens
video head life. The standard emulsion
uses a ferric, or iron, oxide.

The ribbon base is manufactured in several thicknesses. The thinner the base
stock, the more tape can be wound on a
given sized reel. The more tape on the
reel, the longer it can record or play at a
given tape speed. Ribbon tape base is made
in huge wide drums or rolls, which are
coated with the emulsion and which are
later sliced accurately into the narrower
widths, called *pancakes*. These are cut
into the standard lengths used for videotape. The wider the tape stock, the more
tracks and spaces between tracks can be
recorded. The wider the track, the more
gain per track.

Videotape used in quadruplex machines

is 2 inches wide on open reels and runs at 7.5 or 15 ips. Helical-scan C-type machines use 1-inch open reels at 9.6 ips. Helical-scan U-matic cassette machines operate at 3.75 ips with ¾-inch-wide tape. EIAJ black-and-white recorders operate at 7.5 ips and use ½-inch tape. Consumer home product VTRs of both the Beta and VHS type use ½-inch tape. Tape speeds for Beta machines are B1, 1.54 ips; B2, 0.79 ips; and B3, 0.53 ips. For VHS machines they are SP, 1.32 ips; LP, 0.66 ips; and EP, 0.45 ips. See Fig. 7–13.

Cable television uses the U-Matic machines almost exclusively, but consumer product format machines are making inroads as their audio and video quality improves. When we talk of *machine format*, we refer to a combination of tape width, tape speed, cassette or reel design, and tape head azimuth. Only tapes of the same format may be recorded or played on the same or similar VTRs.

Videotape is supplied in three types of container, or package: the 14-, the 9-, and the 5.5-inch open reel, used primarily for the SMPTE/EBU type C standard machines employed in industrial video; the cassette, in at least three sizes, used in the U-Matic; and consumer product machines. The U-Matic cassette is larger and more rugged than the consumer model and employs a thicker ribbon base than the Beta and VHS consumer products (Fig. 7–14). Finally, the 11.75-inch open reel is used on quadruplex machines.

Videotape Care

There is a degree of fragility inherent in videotape. Its cassettes and open reels should be stored vertically to prevent tape edge damage and should be "tails out" (unrewound), which allows the tape wind to become firm, with any storage slack removed, during the rewind before play. Tapes should be stored in about the same relative humidity and temperature that are comfortable for people and should *never* be close to stray magnetic fields. Thus, to avoid partial inadvertent erasure, tapes should never be placed on loudspeakers or any electronic equipment. They should be kept in their protective boxes to avoid accumulating dust, which will be deposited in turn inside the VTR and cause damage there.

Print-through, or *crosstalk,* between adjacent tape layers is the leakage of magnetic energy from tape layer to tape layer. It is most apt to occur in high-level recordings and usually during the first few minutes after recording when the tape magnetic field is its strongest. Crosstalk may also occur at any time when tape layers are too tightly or unevenly wound, as often occurs during the rewind, or if a recording has been stored for a long time without replay.

Finally, any extensive holding of a videotape in pause mode on a recorder or playback will cause wear on both the tape and the head. Leaving a tape in the recorder will cause tape stretch on the tape wound around the head drum.

Videotape Erasure and Storage

Videotape may be deliberately erased, or degaussed, and reused. Video tape recorders have an erase head which is activated in the record mode. This head is located so that the tape is erased immediately before recording.

A phenomenon called *residual magnetism* causes some small amount of magnetic energy to remain on the tape which appears as "snow," or tape noise, when a tape is degaussed on a VTR. So

Recorder	Tape Speed	3M Tape Type	Tape Width	Standard Tape Lengths Offered	Reel Part Number	Maximum Reel size	Approximate Playing Time (minutes)
U-Matic* Recorders Sony, Panasonic JVC, NEC, Concord, Wollensak, Sony LVO-7000 (2 hour U-Matic)	3.75 1.88	UCA-5 UCA-10 UCA-20 UCA-30 UCA-60 UCA-75 MBU-30 MBU-45	¾"	112' 237' 424' 611' 1175' 1453' 611' 850'	Videocassette	- - -	5 10 20 30 60 75 30 45
Mini U-Matic* Recorders — Sony, JVC, Panasonic	3.75	UCA-10S UCA-20S MBU-5S MBU-18S	¾"	237' 424' 112' 345'	Mini Videocassette	- - -	10 20 5 18
Beta Format VCRs Sony, Sanyo, Zenith, Toshiba and others	Beta I 1.57 Beta II 0.78 Beta III 0.52	L-250 L-500 L-750	½"	260' 495' 731'	Videocassette	- - -	30/60 60/120 180/270
VHS Format VCR Panasonic, JVC, RCA, Magnavox, AKAI and others	S.P. 1.31 L.P. 0.66 EP or SLP 0.44	T-30 T-60 T-120	½"	213' 420' 813'	Videocassette		30/60 60/120 120/240/360
AKAI VTS-110 VT-700	11.25	361 or 461	¼"	1200'	R154	5"	20
AKAI VTS-150	10	455	¼"	1200'	R154	5"	26
EIAJ Standard ½" Recorders Sony, Panasonic, Shibaden (Portables usually are limited to reels 5-⅛" or less in diameter)	7.5 7.5	361 or 461 363	½" ½"	845' 1200' 2400' 1550' 3000'	R130 R130 R148 R153 R148 R153	4-⅝" 4-⅝" 5-⅛" 7-⅛" 5-⅛" 7-⅛"	10 20 30 60 40 80
Sony 8650 Series or Panasonic NV-3160	7.5	455	½"	1200' 2400'	R148 R153	5-⅛" 7-⅛"	30 60
Ampex 1" VR-7000 Series VR-5000 Series	9.6	360 or 461	1"	1500' 3000'	R139 R97	8" 9-¾"	30 60
Ampex 4900	9.6	461	1"	3000'	R97	9-¾"	60
Sony EV-200, EV-300	7.8	360 or 461	1"	1250' 2460'	R91 R91	8"	30 60
IVC 800 Series IVC 900 Series	6.91	361 or 461	1"	1075' 2150' 4300' 5400' 7500'	R139 R139 R159 R159 R159	8" 8" 12-½" 12-½" 12-½"	30 60 120 150 210
Ampex VR-1500 VR-660B, VR-1560	3.7	360B	2"	1200' 2400'	R143B R143B	8" 10-½"	60 130
IVC 1010 Series	6.91	479	1"	2150'	R139	8"	60
IVC 9000 Series	8	371	2"	2640' 3900' 4900'	R150 R150 R150	10-½" 10-½" 10-½"	60 90 120

SMPTE Type A or C
Ampex VPR-1, VPR-2,
Sony BVH-1000
  & 1100,
Marconi MR1 & MR2,
Hitachi HR-200, 300
NEC TT-7000, RCA
  TH100, TH200
	9.6	479	1"	1630'
				3170'
				4610'
				**7500'
				**9100'
					R172
					R172
					R172
					R172
					R172
						9"
						9"
						10-½"
						12-½"
						14"
							34
							66
							96
							155
							188

Portable Recorders RCA-TH-50 Ampex-VPR-20 BVH-500, Hitachi HR-100	9.6	479	1"	1630' 3170'	R172 R172	9" 9"	34 66
SMPTE Type B Bosch-Fernseh, Philips, IVC, BCN-40, BCN-50 BCN-51	9.6	479B	1"	1630' 3170' 4610' **6800'	R172 R172 R172 R174	9" 9" 10-½" 12"	34 66 96 140
Portable Recorders BCN-20	9.6	479B	1"	1630' 3170'	R174 R174	9" 9"	34 66

*U-Matic is a registered trademark of Sony Corporation
**Extended Play VTRS

Figure 7–13 Videotape selection guide. *(Provided courtesy of 3M Company.)*

occasional "bulk" erasure is recommended.

There are two types of bulk eraser. One is hand-held and sits on the tape cassette or reel; the other sits on a table on which the tape is placed. In either case, the method is the same. Put the cassette (or reel) on or under the device, after first removing your wristwatch and placing it a distance away. Turn on the degausser

Figure 7–14 U-Matic Cassettes. *(Provided courtesy of 3M Company.)*

power. Rotate the cassette three or four revolutions; then slide it out and away from the degausser's magnetic field to arm's length and put it down. Turn off the power. Turn the cassette over on its other side, and repeat the procedure. Tip the cassette up with its open side facing the degausser, and repeat the procedure.

REVIEW QUESTIONS

1. Compare the quadruplex VTR with the helical-scan VTR.

2. What is a time-base corrector? A sync generator?

3. What type of machine and format is used in cable TV? In broadcast TV?

4. Describe the uses of a VTR or VCR in television.

5. Describe videotape. What is the relationship between tape thickness and running time?

6. Discuss care and storage of videotape.

7. What is print-through?

8. Describe the bulk erasure of tape.

Chapter 8

Video Field Production

The video field shoot, or production, or out-of-studio recording or telecast, falls into one of two categories: electronic news gathering (ENG) or electronic field production (EFP). We treat them separately because each seeks to achieve a different pictorial end, even though in essence they are the same.

ELECTRONIC NEWS GATHERING (ENG)

Sometimes ENG is done for an immediate line feed of the program material to the TV operating facility as a news cut-in and is sent over lines especially installed for the occasion or transmitted by microwave from the TV pickup van. This very expensive procedure is reserved for presidential statements or other similar news opportunities at that high level of importance.

Much more often, ENG is videotaped on site for later editing into a newscast. ENG is the least demanding type of program pickup from an aesthetic point of view and the most demanding type in terms of the physical energy required of the camera crew. The ENG endeavors of an access cable TV crew will undoubtedly

be at a much less demanding level than those of a broadcast network crew; but by describing the operations of the network crew, we encompass the operations of the access crew.

An ENG pickup often requires racing to the action scene (a vehicle crash or a bank holdup in progress) in a van or other vehicle and setting up camera and tape equipment under adverse weather and lighting conditions. Then the crew must shoot the newsworthy tape footage and return to home base, perhaps editing the tape in the van (if the vehicle is so equipped) during the return trip.

More often, ENG involves a crew being scheduled in advance to cover a speech or press conference. It means they must arrive at the scene early enough to set up lights and equipment, arrange for power, get a printed copy of the speech or advance statement, get permission to set up and tape, and perhaps, if the event is a luncheon or dinner speech, wangle an invitation for the crew to join in the meal. Getting there early concerns more than simple worry about road traffic conditions. It means beating the competition to the best camera and microphone placement spots at the site.

ENG coverage requires a great deal of

advance equipment-use planning. The minimum equipment taken along includes a video camera, which may be used on a tripod or hand-held, including a choice of zoom lens systems; a portable lights package; a VTR with more than enough videotape cassettes; and an audio mixer and microphones. In addition, there must be camera-to-VTR cables and extension cables, mike cables and extension cables (or a wireless mike system), and power cables and extension cables (or battery packs and spare, fully charged batteries).

The ENG Coverage Crew

The coverage crew must include someone to set up lights; someone to handle the audio, including riding the audio gain and operating the VTR; someone to operate the camera; and, of course, a driver. If the newsworthy event is an ongoing one, a messenger may be needed to ferry tapes from the site back to the TV facility.

In the worst-case situation, one person does all these things; in the best-case situation, a separate person does each task. Also a camera director may be assigned to the crew; but if there is not, the camera-operator directs the crew. In the most usual situation, the cameraman also does the lighting, and the audioman also does the videotaping. Where there is one, the third crew member is usually the lighting technician. In all cases, the crew works closely together, and each person assists in all functions.

The ENG coverage crew is assigned to a story by a coverage editor in the news department, and the crew can be pulled off that story, even in the midst of shooting, if something bigger or more newsworthy breaks. This often happens in re-

sponse to a telephone page receiver or a two-way walkie-talkie carried by the crew.

A typical ENG assignment might be coverage of a scheduled event, perhaps an interview in which only the person being interviewed is seen on camera, responding to off-camera questions. Alternately, it might be a "stake-out," in which the crew waits, with the equipment set up, for someone important who is participating in a meeting to emerge from the meeting and read a prepared statement and perhaps answer some questions. The footage that is shot should always include cutaways. These are described later.

ELECTRONIC FIELD PRODUCTION (EFP)

Electronic field production involves use of a full complement of transportable (not necessarily portable) studio/control room equipment. That is, more than one camera is used, and those cameras are fed through a switcher.

The equipment is set up outdoors in order to make full use of live background scenery, which either could not be duplicated in a studio or would be more costly to duplicate than traveling to location shooting. The aesthetic values are the same as in studio shooting. All production parameters are planned and carefully executed or reshot with additional takes if necessary.

The program is scripted, and both talent action and camera shots are blocked out in advance. Audio is taped along with video, but may be replaced, that is, *looped*, or dubbed in, during postproduction either because of presence problems in an outdoors, nonreflective audio medium or because of extraneous background noise which must be eliminated.

Sometimes EFP techniques are used to

produce unscripted coverage of man-in-the-street interviews and other documentary program segments. Since this material is taped with editing in mind, it is vital to also tape *cutaways*. These are reverse-angle shots, side-angle shots, and others in which there is no direct, head-on picture of the person speaking (that is, the speaking mouth is not seen).

Cutaways are used in the editing process to cover *jump shots*, which are two shots that are almost identical but different enough that the picture seems to shift (jump) vertically, horizontally, or both. A jump would occur, for instance, if an edit were made of someone sitting and speaking. If the edit came as the speaker shifted her weight from the left to the right side in the chair, then she would seem to jump horizontally from left to right, perhaps only a few inches, when the edit was completed. The jump is avoided by interjecting the cutaway at the edit point. So the subject is seen talking and leaning left; then the camera cuts away momentarily to the back of her head, facing the audience, and again cuts back to the original shot, but with the speaker leaning right this time.

The EFP camera operator should always edit mentally while shooting. His thinking should always encompass the final product. For example, when covering a long speech with a marked-up advance copy of the speech in hand, the camera operator should tape only those parts that are relevant to the story or the program. Just as it would be foolhardy to undershoot and have insufficient relevant material from which to select the final product, so would overshooting mean additional precious time spent in screening irrelevant material.

Both ENG and EFP production use the same camera equipment as we have discussed previously, with the exception being the VTR, which in this case is a portable model. Now we look at the Sony VO-4800 as an example of one portable VTR used in cable television (Fig. 8–1).

SONY VO-4800 PORTABLE U-MATIC VCR

The Sony VO-4800 is a portable U-Matic video cassette recorder. It will record for up to 2.5 hours on its self-contained battery or up to 1 hour when the same battery is powering a portable color camera. It is considered good practice, however, to power both the camera and the VTR, each with its own supply. The VO-4800 is a top-loading recorder whose loading control EJECT is the leftmost control on the front of the machine. This control is followed, left to right, by the fast-forward (F FWD), PLAY, rewind (REW), record (REC), and audio dub controls.

Above these piano key control switches is the STOP button and the PAUSE control, together with a green LED which blinks while the recorder is in the pause mode. The pause mode automatically switches to the stop mode after 8 minutes.

Figure 8–1 Sony VO-4800 VCR. *(Provided courtesy of Sony.)*

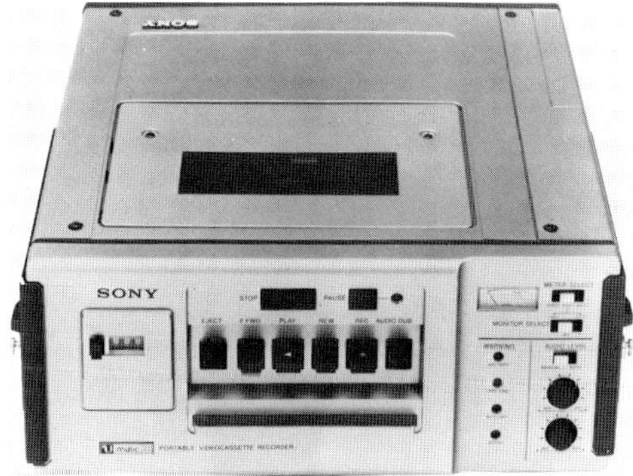

Although both audio and video levels may be controlled automatically, it is best to control audio manually. A controls section to the right of the mode controls just described permits metering of either battery, audio or video; monitor selection of Channel 1 or Channel 2 audio; and a manual/automatic select switch and two pots, to control audio input. There are also warning LED lamps for low battery; for tape end, which signals that the tape is within 2 minutes of its end; for automatic off (AUTO OFF), which illuminates when the recorder stops because of moisture on the head drum or loose tension on the tape; and for servo, which blinks if either the drum servo or the capstan servo fails to lock.

The right side of the machine has the input and output connectors necessary for feeding video and audio signals in and out of the recorder as well as a battery compartment for internally powering the machine.

The VO-4800 weighs less than 22 pounds with battery and cassette installed. It uses KCS 10 and KCS 20 U-Matic cassettes.

The specifications sheet shows these data:

1. Video
   a. Input: Composite, sync negative, 1 V peak to peak, 75 Ω unbalanced
   b. Output: Composite, sync negative, $\pm$ 2 V peak to peak, 75 Ω unbalanced
   c. Horizontal resolution: 260+ lines
   d. S/N ratio: better than 45 db
2. Audio
   a. Input: Mike: -60 db, low Z
      Line: -10 db, $Z = 100$ kΩ
   b. Output: Line: -5 db at 47 kΩ unbalanced
   c. S/N ratio: better than 48 db
   d. Frequency response: 50 Hz to 12 kHz

3. Tape speed: 3.75 ips
4. Wow and flutter: 0.25% root mean square (rms)

ENG PRODUCTION PROCESS

Let us look now at the processes involved in an ENG production. The ENG crew is scheduled to tape a 3 p.m. news conference. Travel time and setup time together are estimated to be 1.5 hours. The crew leaves for the shoot at 12 noon, three hours before the event is scheduled. Upon arrival, it is discovered that other crews are there and will be taping also. A quick agreement is reached in a friendly fashion as to how much lighting is needed; and the lighting setup task is shared among the crews, because they all will use the same light.

Tripods for cameras are set up, as are mike stands on the podium or at the spot where the speaker will stand (the earliest arrivals get the best positions). If there will be more than one speaker and if the speakers will be seated at a table, then a mike for each audio setup is placed in front of each speaking position. Often in the larger cities of the United States, an unsightly battery of mikes in front of a speaker is avoided by the assignment from one of the TV stations of a mike "mult"— a unity-gain amplifier with many mike level outputs and two mike inputs—together with an operator for the mike mult. Then each audio user, including radio technicians, take feeds from the audio mult.

Mike cables are dressed carefully away from foot traffic, to the mixer, to the mult, or to the VTR.

The camera is placed on its tripod. The battery or ac/dc power supply is attached and turned on. The camera power switch is turned on and switched from standby power to full power. The camera is set up and is white- and black-balanced, and the lens is zoomed tight and focused on a stand-in for the speaker. The camera is ready to shoot.

The VTR is set up next to, behind, or near the camera. The multipin connector cable is connected between the camera and the VTR. The VTR operator gets a tentative audio level from the stand-in, just to ascertain operation of the mike(s) and his earphones. The VTR controls are placed in PLAY/REC, which is the record mode, with the tape running. If the record control alone is depressed, the record signal will be displayed on the monitors, viewfinder, and VTR meter during the pause or freezeframe mode.

The cameraoperator switches from video to color bars on the camera and feeds color bars to the VTR for 30 seconds. During that time the VTR operator observes video levels and, using a mike, makes an audio identification of the tape, including date, subject, names of crew members, and tape number. Then the camera is switched back from color bars to video output, the VTR is turned off, and the camera power is switched to standby.

The crew then relaxes until the program starts. During this interval, arrange-ments are made for procuring advance copies of statements and any other printed material that would be useful in coverage of the story.

At the conclusion of the videotaping, the camera is carefully turned off, discon-nected from its cables and tripod, and re-placed in its protective case. Mikes are repacked, cables are carefully coiled, and every piece of equipment is made ready for the next shoot. A battery-use log is often maintained to ensure that batteries are recharged when needed. The crew re-turns to home base, and the tapes are de-livered to the tape editor.

REVIEW QUESTIONS

1. Define the acronyms ENG and EFP.
2. Why is it wise for the crew to get to the shoot early?
3. What equipment does an ENG crew carry?
4. Why are color bars put at the front end of a videotape? What else is put there?
5. Describe cutaways. What are they used for? What is a jump shot?
6. Why should one crew remain on good terms with another crew of the "op-position"?

Chapter 9

Video Editing and Postproduction

A television program is called a *production* regardless of whether it is short or long—a 30-second "spot" commercial is a production—and it is created by a program producer.

Productions are created as artistic endeavors, much as paintings or still photographs are. Either productions are accomplished live, in which case all the action takes place in real time (that is, the production is shown to a viewing audience at home as it happens), or they are painstakingly pieced together from prerecorded material in the quiet atmosphere of an editing room.

Live productions include the use of cameras, the action of participants, a lighted studio, video and audio switching with mixes and effects, and perhaps the insertion of prerecorded segments.

However, it is often more expeditious for the producer to accumulate the various recorded elements of a production and edit them in what is called *postproduction*. But this can be done only if the program has no instantaneous demands, such as a press conference, which would need to be shown immediately if it were not to suffer a complete change in value

to its audience. If a press conference which revealed decisions of immediate importance were delayed, it would change in value—from news to history.

Postproduction requires the skills of a videotape editor, a person possessing an infinite amount of patience, because all the prerecorded elements of the production must be individually and tastefully judged for viewer interest, accurately timed, and then carefully edited to form the finished product.

Audio must be carefully intermixed with the video. And both the audio and video edits that are less than perfect, either technically or aesthetically, must be repeated until near perfection is achieved. Most of all, postproduction can be very time-consuming, especially for someone who does it only occasionally.

In the early days of videotape editing on open-reel quadruplex machines, the videotape was actually cut with a razor knife at editing points and spliced together, much as audio tape is still edited. The vertical-interval edit point was found by moving a glass disk containing iron filings across the tape and observing as the tape's magnetism displaced the filings in a crude display of the vertical interval, and thus the edit point.

In current practice, videotape that is edited for postproduction is dubbed, or copied electronically, from one tape to another at the required edit points.

VIDEO EDITING TOOLS

The tools of postproduction in television usually include the following:

1. One or two videotape *playback* machines.
2. A time-base corrector for each videotape playback machine.
3. A video tape *recorder*.
4. An editing controller.
5. Two or more video monitors, one for each tape machine.

Sometimes a time code generator and a time code reader are needed in the editing process, if time code is used instead of the pulse code generated by the VTR. We discuss this more later.

Helical-scan VTRs using either ½-, ¾-, or 1-inch tape can be employed in the editing process. Often, ½- or ¾-inch tapes are "bumped up" to 1 inch or re-recorded at 1 inch before editing.

Freeze Frame

As described in Chapter 7, helical-scan machines have a rotating-head cylinder, or wheel. The rotating wheel's two heads write during recording, or pick up magnetic impulses in the playback mode, a complete field of video during each rotation of the wheel. If the tape is not moving but the head wheel is, then the picture output of that machine is a *still frame*—one frame of picture, frozen in time, and thus often called a *freeze-frame*. Recall,

too, that although the video information is written diagonally on the tape, the audio tracks contain information written horizontally and by a nonmoving head.

An output of the video playback machine, either audio or video, requires that there be *relative* movement between the tape and the playback head. Therefore, when the tape on a video playback is not in motion, there is *no* audio but there is freeze-frame video, because of the motion of the head wheel, which produces the relative tape-to-head motion.

Control Track

Depending on the make of VTR, up to three horizontal "audio" tracks may be available on the videotape. There are always at least two horizontal tracks, and one additional horizontal track will be generated by the recorder as a *control* track. The control-track signal, a pulse for each frame of video, is generated internally to serve as a timing standard against the ac line frequency; but it serves as an editing aid as well.

The *editing controller* is an electronic device for controlling the speed, direction, and operational modes of two VTRs. One of those controlled recorders will always be in the record mode during the editing process, and the other will be in the playback mode. Editing controllers vary in complexity, and therefore in price, by the number of playback VTRs that they are able to control. Here we discuss editing controllers that have the capability of controlling one playback.

TIME-BASE CORRECTION

Because they are mechanical-electronic devices, helical-scan VTRs, from the simplest to the most complex, have a built-in problem called *time-base error*. This is another way of saying that the VTR output is somewhat unstable horizontally, or that the output contains *jitter*.

If the VTR output is connected directly to a video monitor, the jitter is not apparent to the viewer because the stabilizing circuits of the monitor lock onto the jittering signal and follow, rather than display the jitter. If the monitor stabilizing circuits failed to lock onto the VTR output, then some of the picture lines would be shifted to the right and others to the left, in typical horizontal distortion of the picture, displaying the time-base error in the recorder's horizontal video signal.

The device and the electronic circuitry used to correct this error are called a *time-base corrector*. It is a computer that processes each horizontal line of video and resets the timing of that line to exactly match the standard time base for a horizontal line of 63.5 microseconds.

Now, if an attempt were made to mix the *direct* signal outputs, without time-base correction, of two or more VTRs, a serious problem would arise: Each VTR has its own time-base error, and a constantly varying error at that. The two machines would have independent errors that never match, and it would be impossible to accurately combine their signal outputs.

One method of mixing nonrecorder output sources with the output of a VTR playback is to make all the other sources of video follow the jittering signal of that one playback VTR. To do this, it is necessary to have a sync generator that can follow the time base error of a VTR and to genlock that sync generator to the VTR. This method works only with signal sources that can be pulled to follow a time-base error, and not all sources can perform

in that manner. It is often possible to mix the output of a VTR and the output of a camera, depending on the make and model of the camera, but *not* between two VTRs.

To mix VTR outputs, each output must feed the mixing device through a time-base corrector.

There is another consideration regarding the above mixing method: In locking other video sources onto a jittering signal, the composite signal that is the output of the mix will also be a jittering signal. If this signal is recorded, the second generation of dub, or copy, will have twice as much jitter as the original tape output since the jitter of the second machine is added to that of the first. Of course, further generations of tape copy get worse, and the editing process is based on repeated copying.

It is possible to remove jitter from a second- or third-generation tape by processing it through a time-base corrector, but some jitter residue will be left, because of the complexity of the signal at that point. Therefore, it makes more sense to remove jitter at its source by using a time-base corrector at the output of each playback VTR.

Even though it is not possible to *mix* the outputs of two tape machines without time-base correctors, it is possible to perform *cut edits*, which are edits during the vertical interval, because most playback machines can be made to lock "vertically" onto each other. Even though there are horizontal and even color shifts between the two machine outputs, the machines can be made to perform a vertical-interval switch, or cut, between the signals of the two machines.

This process works for cuts but not mixes, because during a mix the pictures are integrated line by line, every line, from top to bottom of the picture, and a cut is a one-time occurrence during the vertical interval. If the cut is made other that at the vertical interval, the picture will "roll" at the point of the cut and can be seen on the monitor during each playback.

Further, remember that there are two fields to a frame. One field is composed of all the odd lines, and the other, following field is composed of all the even lines. When videotape is edited, it is necessary to ensure that all the recorded material has the same sequence of odd field, even field. If two odd fields, or two even fields, follow each other, serious video distortion problems occur. Similarly, correct color framing must be maintained for a clean edit.

The control-track pulses recorded on the videotape are used in editing to *log*, or reference, recorded material by frames. One pulse is recorded for every frame of video. By counting the number of pulses and keeping track of the direction of tape travel, it is possible for the recorder and playback to follow tape movement back and forth during edit searches and then to use the control pulses as a reference to establish edit points on the tape.

During the edit search, the tape editor is looking for *beginnings and endings* of tape segments to be included in or excluded from the production.

Problems with this pulse method of tape editing begin when the tape is removed from the machine, as it often must be, in order to look at other tape segments, and then replaced. If the tape in the cassette or on the reel is moved even slightly while it is out of the machine, then all pulse reference to that tape is lost for that particular editing sequence. If the editing pulse reference is lost, it must be redetermined, and new reference points need to be programmed into the machines.

Additionally, it is easy for the editing controller to miss control pulses when a tape's speed is changed or its direction of travel is reversed.

TIME CODES

These concerns have led to the use, by videotape editors, of either one of two special codes for keeping track of specific locations on a videotape.

One code is called the Society of Motion Picture and Television Engineers (SMPTE) code. It is also called the European Broadcast Union (EBU) code. This code, generated by an external or internal time code generator, employs one of the tape's audio tracks (other than the one on which the internally generated pulses are recorded). The SMPTE code assigns a specific number to each frame, and that number is recorded digitally on the track. When the videotape is edited, an SMPTE time code reader is used to keep track of edit points, even at fast-forward and fast-reverse speeds. The SMPTE time code reader loses the code at very slow tape speeds, but picks it up exactly when the tape is speeded up.

Another system of tape control is called

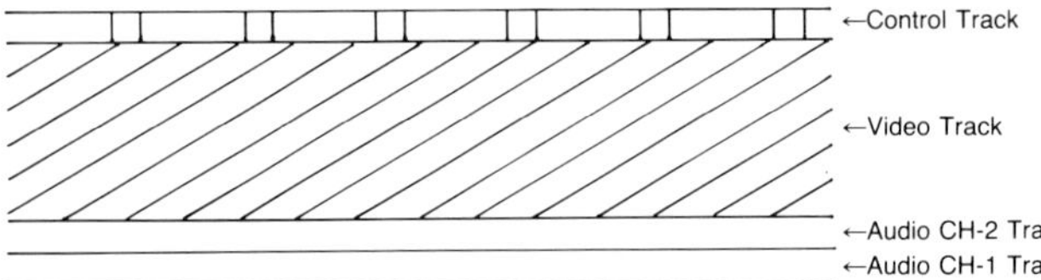

Figure 9–1 Videotape format, helical scan.

Micro Lok. This system modifies the original control-track pulse signal by shortening every hundredth control pulse, and this shortened pulse is used to reset the frame counters on the editing controller. Even if a few pulses are lost in moving the tape back and forth from fast forward to fast rewind while looking for edit points, the tape frame counters are reset each time that a short pulse is encountered and recognized. The assumption is made that no more than 100 pulses will be lost in moving the tape back and forth while looking for edit points. However, to get absolute accuracy in positioning, the tape must be rewound to a start point, to get the correct data, and then moved forward to the edit point.

Now that we know what hardware to use, let us describe how to make an edit.

THE SIMPLE CUT EDIT

With equipment arrangement shown in Fig. 9–2, only cuts may be performed and no dissolves or wipes are possible. Sec-

tions of one tape may be rearranged in different sequence, or parts of many tapes may be added, one cut at a time, to create a composite tape. Audio may or may not be added along with video.

In Fig. 9–2, one playback machine feeds the audio and video signal to one video recorder, and each machine has a monitor on its output. An editing controller is connected between, and controls, both machines. The edit is performed in this instance by using control-track pulses to find edit points.

Insert the playback tape into the playback machine; and press either the play or rewind button, depending on where the tape is on its reel, to load the tape cassette into the machine. While you are playing the tape, adjust the VTR tracking control for maximum head-to-tape contact. If necessary, adjust the skew control to correct distortion (horizontal tearing) at the top or the bottom of the picture. In most cases, however, the set point, or detent, on the tracking and skew controls is the correct setting for properly recorded tapes.

Figure 9–2 Simple cut edit equipment setup.

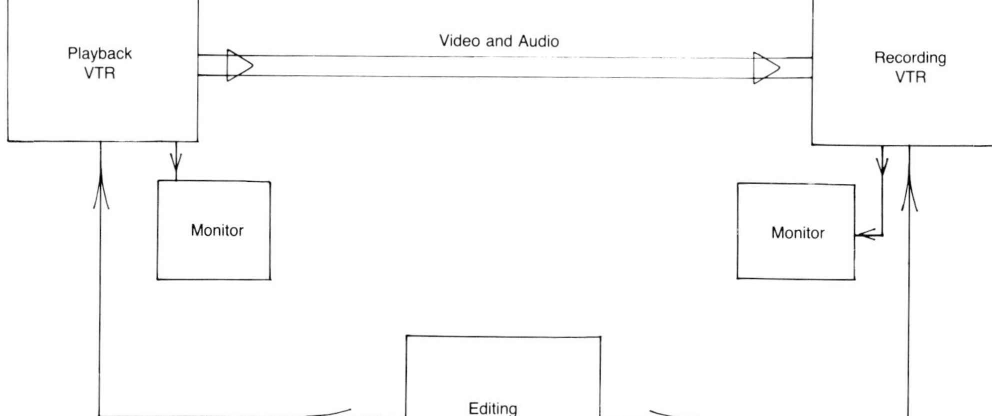

Insert the record tape in the record machine; with the playback tape rolling, put the record machine in the stop mode, with the record (REC) button depressed in order to "see" the playback machine. This is called *E to E*. Using the signal coming from the playback, set the record machine's input video level, with the control set for manual. Then switch the level control to automatic gain control (AGC), and leave the control in AGC during editing.

Adjust the audio level of the record machine, on the channel used for audio.

ASSEMBLE AND INSERT EDITING

Two modes of editing can be used: assemble (or assembly) editing and insert editing.

In *assemble editing*, the video and audio signals are edited (recorded) at the same time, and the *control-track signal is also edited at this time.* Because the edit is performed diagonally on the tape, there is one point at the edge of the diagonal where there is very little signal, and the picture may be somewhat unstable at that point. In assemble mode, tape pieces are butted together until the recording is assembled. With each assembly addition, new control track is added, possibly causing some control discontinuity. Since this mode is less stable, it should be avoided except for the simplest "scratch" edits, or "workprint" assembly.

In *insert editing*, which is the preferred editing mode, the video and the audio channels can be edited either independently or at the same time. Signals are edited by using the control-track pulses previously recorded on the tape as reference. Insert editing is used to add new scenes to a previously recorded tape, to add or change audio to previously recorded video, or to add or change video to

a tape with previously recorded audio. For insert editing, the control track must be *on the tape* before editing begins. This may be done by recording the output of a capped video camera (with or without going through a switcher), thus recording sync information and the reference black signal; by recording the output of a standard sync generator for at least 5 seconds before preroll time and running until the end time of the tape segment; or by dubbing a special videotape containing color bars, a countdown signal, black and sync signal with no discontinuities, and, of course, control track.

Since the control track has been prerecorded, the edits will be more stable (cleaner), and the chances of the tape rolling at the edit point will be minimized.

THE EDITING PROCESS

Now we start the editing process. Since both machines must be brought up to normal speed before they can produce a viable picture, both are "prerolled" by the editing controller before they get to the point on the tape where the edit is to occur, normally just a few seconds from the point where the tape is rolled. During preroll, the tape editor uses the editing controller speed control knobs to ensure that both tapes arrive at the edit point simultaneously. This is done by watching both pulse counters on the editing controller, so that the number of pulses away from edit remains the same on both playback and record machines, and adjusting the speed knobs on the editing controller to keep that parity.

During the edit search, when the exact edit point (either the beginning or the end of an edit) is determined, it is "marked" in the editing controller memory by depressing entry buttons (EDIT IN or EDIT OUT) for both the playback and record machines. The memorized edit-in and edit-out points can even be adjusted frame by frame, by using TRIM buttons to move the tape.

Just before the edit, both machines are in playback mode. The record machine gets an input signal from the playback machine, which it uses for the purposes of "locking up." At the edit point, the record machine changes from the play-back mode to the record mode and starts to record the signal to which it has just locked up. Note that the record machine preerases all prior signal which was occupying space on the area of the tape on which it is now recording.

Just prior to the record command from the editing controller, the recording machine is preset to perform either a video or an audio cut edit or a video and audio cut edit simultaneously. At the instant that the machine is to switch from play-back mode to record mode, it receives a *record* command pulse, and it obeys that command. Most editing systems are not "frame-accurate." The exact edit point may be plus or minus nine frames away from the selected edit point, which translates to about a ⅓-second error factor on each edit. When you select edit points, therefore, allow for that possibility by selecting audio edit points with audio pause preceding and following.

Let us look at an example of an edit. There is a segment of tape called "U.S. Capitol" on the record machine and a segment called "Senate chamber" on the playback machine. Let us assume that we want to keep the first half of the segment "U.S. Capitol" and to replace the second half of "U.S. Capitol" with the second half of "Senate chamber," which is on the playback machine.

After we select the edit points, we au-tomatically preroll both machines for a few seconds, so that they both arrive at the center point of the two segments at precisely the same instant. At that point, the record machine goes into record mode and starts recording whatever is fed to it from the playback machine, which in this case is the second half of "Senate chamber." And since the playback machine never goes into record or erase, the original source material remains intact and may be used should the edit need to be repeated.

More complex edits are possible if the tape playbacks feed through a time-base corrector, if all devices are genlocked to a sync generator, and if a switcher is available. All the signals, then, are time-synchronized to a single source (the sync generator). And edits can be made from a variety of sources, which can be switched, dissolved, or routed through a special-effects generator, either manually or automatically. (See Fig. 9–3.)

In complex edits, more than one play-back machine is controlled by the editing controller. All machines are made to pre-roll and arrive at the edit point at the same instant. Each playback machine has a time-base corrector, and their output signals are fed to a video switcher which can do mixes and add effects.

Adding to our previous tape illustration, we assume that there is a segment "House chamber" on a second playback machine and that we want to do a dissolve from the first playback machine to the second playback machine at the instant of edit. The switcher must have at least two buses—the "from," or *program*, bus and the "to," or *preview*, bus. The program bus signal is the signal that the switcher is presently feeding to its output; the preview bus signal is the signal that the switcher will be wiping or mixing to.

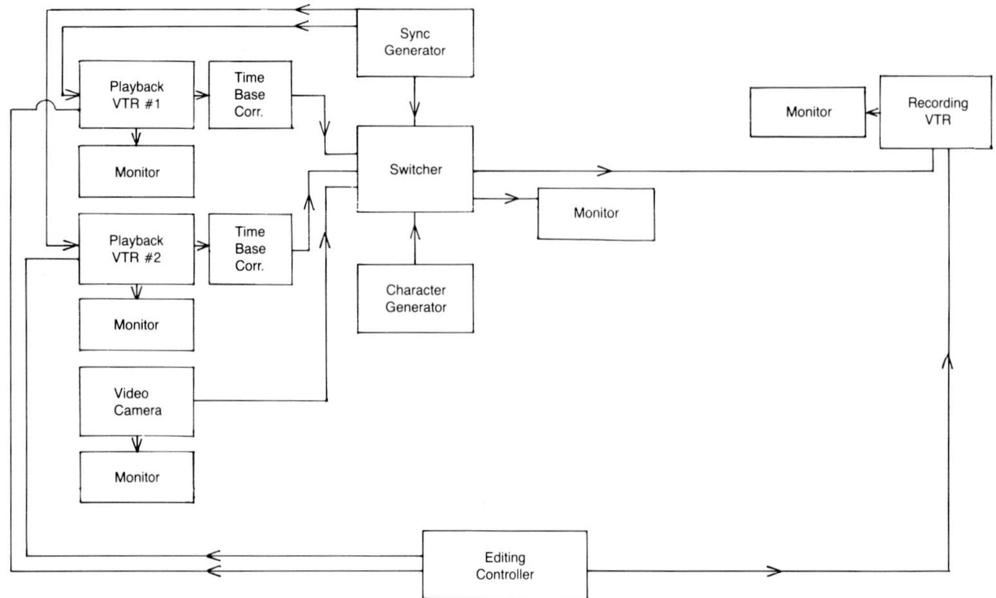

Figure 9–3 Equipment setup for complex edits.

The output signal of the switcher is feeding into the record machine.

As the three machines preroll, the record machine is being fed the signal from the first playback machine, and it locks up to this machine. The two time-base correctors are locked to each other, so that their signal outputs are exactly in phase.

As all three machines come up to the edit point, the switcher is still outputting the signal of the first playback, the "Senate chamber" machine. At the edit point, the record machine starts recording the signal being fed to it. At the same instant, the switcher is commanded to do an automatic mix, and it proceeds to mix the signal from "Senate chamber" to "House chamber," line for line, until the mix is completed and the switcher is outputting the signal from the second ("House chamber") playback into the recorder.

Let us now examine the edit just performed, and we look at it on what was the record machine: First the record machine is switched to rewind mode, and the tape is rewound to the beginning of segment "U.S. Capitol." Then the machine is put in playback mode and the tape is run, while we watch the monitor. The first part is segment "U.S. Capitol," and at a point, the signal will switch to "Senate chamber," hopefully cleanly. Then it will immediately start a dissolve to "House chamber" and run to the end of segment "House chamber."

Unfortunately, we saw very little of segment "Senate chamber" because the mix started so soon after the "Senate chamber" beginning, and this clearly would not be a good edit for our program.

It would have been more effective programmatically to have seen more of "Senate chamber." The edit should have been done with a *delayed transition*, or delayed effect. The effect can be delayed in one of two ways: The first way is to simply do two edits. The first edit would cut from "U.S. Capitol" to "Senate chamber." The second edit, performed a few seconds farther along the tape, would include the mix from "Senate chamber" to "House chamber."

Another faster and more professional way to make the delayed transition is to delay the transition command to the switcher for a few seconds after the edit point; and most sophisticated editing controllers can do a delayed command. Here, then, is the sequence of events with delayed command: All three machines start on preroll, and at the edit point the record machine switches to record mode and starts recording the switcher output, which is "Senate chamber" playback. Then, a few seconds later, the switcher receives the transition command pulse which starts the mix from "Senate chamber" to "House chamber." The mixer completes the transition to "House chamber" and continues until the end of the "House chamber" segment.

We have virtually ignored audio as part of the editing process up to this point, and we make haste to include it here. The videotape recording may have audio on one or more tracks. Most often it has audio on only one track, usually track 2, while the second audio track may contain SMPTE time code. Generally, when the tape is made in the field, only one audio track is recorded, and there is no recorded time code because this requires carrying additional equipment. Audio *can* be recorded on two tracks, and it may be stereo audio or perhaps bilingual audio.

A simple system for creating a more complex audio track, perhaps with narrator, music, and effects, can be done by installing an audio mixer between VTR playback OUT and VTR recorder IN. If the narration has been recorded on VTR channel 2, then later you should record appropriate music on VTR channel 1. Finally, when feeding audio from the VTR playback to the VTR recorder, you mix the two audio channels through the audio mixer for a pleasing audio composite track to the VTR recorder.

Portable time code generators do exist and may be carried when a big field production is anticipated. In any case, time code can always be added to the recorded tape in the editing room before the editing process is begun.

During editing, sometimes the audio is made to follow the video, and at other times it is more convenient to *split-edit*.

Here is an illustration of audio-follow-video. An outdoor scene is dissolved into an indoor scene of two people discussing the outdoor scene and is dissolved again into the outdoor scene. Here it would make sense to have the audio follow the video. First, record audio and video of the complete indoor scene discussion, including the description of the outdoor scene. Later in an insert edit, record *only video* of the outdoor scene over the appropriate dialog, which will remain the same as in the first take.

An illustration of split audio occurs when a studio announcer is shown commenting on an outdoor scene, the studio scene is faded to the outdoor scene with another announcer describing the outdoor scene, and then the outdoor scene is dissolved back into the studio scene, with the studio announcer's voice dissolved back into the commentary. In this instance, the edits are both audio and video edits.

SONY RM 440 EDITING CONTROLLER

The Sony RM 440 Automatic Editing Control Unit is perhaps the most popular editing controller used in cable television (Fig. 9–4). The operating controls on the

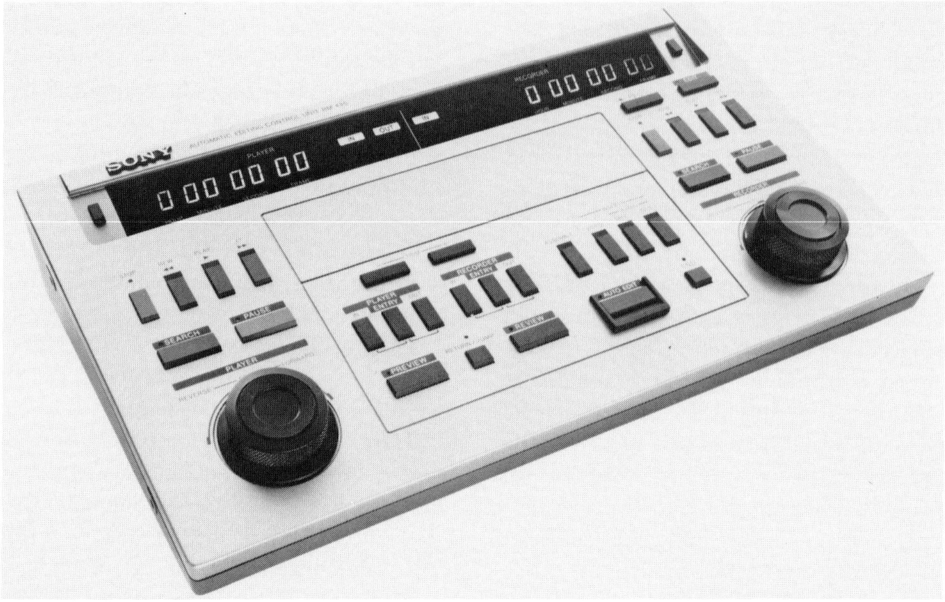

Figure 9–4 Sony RM 440. *(Provided courtesy of Sony.)*

sloping front panel of the RM 440 are divided into three sections. On the left side, the large knob and the pushbuttons above it control one video *playback*. The pushbuttons are STOP, rewind (REW), PLAY, fast-forward (F FWD), and PAUSE, similar to the corresponding controls on the playback machine. Note that there is no record button on this side of the RM 440 controller.

An additional button, the search button, just above the knob or dial, disables or enables the search control knob. The dial controls the playback machine speed from slow to still to high speed; from one-thirtieth of the normal speed to freeze frame to five times the normal speed, in conjunction with a switch on the back of the RM 440.

On the right side, the search knob and the operating buttons above it control the video *recorder*. The buttons are record (REC), STOP, rewind (REW), PLAY, fast-

forward (F FWD), and PAUSE; they, too, duplicate their counterparts on the VTR.

The operating buttons in the center of the panel control automatic editing operations, and the LED readouts across the top, are digital time counters that count the control pulses on the tape and display the results in hours, minutes, seconds, and frames, for each of the two tape machines. There is a timer reset button for each machine, with the playback on the left and the recorder to the right of the readouts.

The automatic edit buttons are the trim buttons, which move the edit-in and edit-out points frame by frame, at the top. Below these are two sets of entry buttons, for edit-in and edit-out on both the playback and record machine.

Below these, left to right, are the preview button, which permits rehearsal of the edit, and the return/jump button, which controls two modes. During the preview mode, the recorder will return at high speed to the edit-in point and remain in freeze-frame. During the review mode (discussed next), after the edit-in point, the return/jump button fast-forwards the just edited material and permits monitoring of the *edited* picture from 5 seconds prior to the edit-out point to 2 seconds after the edit-out point. This permits a critical appraisal of the completed edit and is used to save time in the event of a long edited piece. After the edit point is checked, the tape can be speedily moved forward to select the next record edit entry.

The review button permits review of the edited picture. The automatic edit button places the recorder in the automatic editing mode. And the END button, when in preview mode or in AUTO EDIT, will mark the edit-out point and during manual editing will end the edit.

Above the AUTO EDIT and END buttons are the assemble button, which puts the editing controller in the assemble mode, and three input selector buttons for insert editing.

Above this grouping, on the upper right of the editing controller, is the EDIT button. When it is depressed simultaneously with the PLAY button, the operation of the controller shifts from automatic to manual editing.

We talk about the back of the RM 440 to indicate where the 33-pin connector cables are mounted, one to the playback VTR, on the right side, and one to the recorder, on the left side. Between the cable mounts is the cue out connector, which provides the cue signal for camera recording. Also between the mounts are five selector switches (left to right): The reverse switches, one for playback and one for recorder, are set to ON when the VTR has reverse function and set to OFF when the VTR has no reverse function. The search dial switch, set to DIRECT, elim-

inates from the circuit the search enable/disable switches above the front search dials; when set to STEP, it places the enable/disable switches into the circuit. The preroll time switch sets preroll to either 5 or 10 seconds. And the frame switch is set to 30 for NTSC television or to 25 for PAL/SECAM systems.

The digital time counter on the front panel will count, using the NTSC system, from -9 hours 59 minutes 59 seconds 29 frames to $+9$ hours 59 minutes 59 seconds 29 frames.

The editing accuracy without previewing is from ± 3 to ± 5 frames and with previewing from ± 5 to ± 9 frames. Power to the editing controller is supplied from the recording VTR.

Beta (½-inch) tapes as well as U-Matic (¾-inch) tapes may be edited by using the RM 440.

In the assemble editing mode, the video, audio channel 1, and audio channel 2 signals can be edited at the same time. In the insert editing mode, the video, audio channel 1, and audio channel 2 signals are selected independently, and any combination can be selected.

Butt editing, or continuous editing, can be performed by choosing the edit-out point as the next edit-in point, instead of reentering the next record edit-in point.

During editing, the pause mode of the recorder is automatically released after 6 minutes of holding in freeze frame pause, and the tape is forwarded at the minimum search speed, to protect both the video head and the tape from damage.

Assemble Editing with a New (Blank) Tape

It is not usually necessary to record the control signal in advance. But if the assemble edit is made from the beginning of the new tape or after a blank spot on the tape, a control signal has to be recorded for at least 5 seconds (10 seconds if the preroll time switch on the back of the RM 440 is at the L position) prior to the first edit-in point.

Insert Editing on the RM 440

Sony VTRs have either a 20- or a 33-pin remote-control connector on their connector panels.

With a 33-pin connector machine:

1. Set the insert mode switches on the VTR being used as a recorder to the OFF position so that the insert mode can be turned on and off by the RM 440.
2. Select the desired input signal with any of or all the insert buttons on the RM 440. An indicator lamp will light. If the indicator does not light but the assemble indicator is lighted, press the assemble button to disengage assemble, its indicator will go out, and then repress the insert button(s).

With a 20-pin connector machine:

1. The input signal cannot be selected by the RM 440. Select it with the insert mode switches on the VTR being used as a recorder.
2. Press any one of the insert buttons on the RM 440, video, audio channel 1, or audio channel 2. The audio channel 1 indicator will light, indicating that the recorder is in the insert mode.

Video tape recorders that employ remote-control cables with other than 20- or 33-pin connectors will require that special interconnect cables be wired in order to be used with the RM 440 editing controller.

To begin editing, we first determine the beginning, or edit-in point, and the end, or edit-out, point of a scene.

On the Playback Machine

To find the edit-in point:

1. Put the machine in the playback mode by turning the search knob on the left, or playback, side of the RM 440. If no tape motion is observed, press the search button and then turn the dial knob. If the playback has no reverse function, it will not reverse when the knob is turned to the reverse side. Use the REW button to reverse the tape.
2. Locate the point on the tape where the scene is to start, by watching the video monitor. At that edit-in point, turn the search dial to 0, to pause the tape.
3. Simultaneously depress the IN and ENTRY buttons on the RM 440, to enter the edit-in point on the playback machine. The IN indicator lamp will go from blinking to ON, and the LED indicated numbers are memorized as the edit-in point. Note that the first edit-in point should be at least 5 seconds after the beginning of the tape or 10 seconds if the preroll time switch is set to L. If the edit-in point is to be changed, locate the new point and depress the IN and ENTRY buttons again.

To find the edit-out point:

1. Locate the point on the tape where the scene is to end by turning the search dial and watching the monitor. Then set the dial at 0.
2. Simultaneously depress the OUT and ENTRY buttons for the RM 440 to memorize the playback edit-out point. The OUT indicator lamp will cease blinking and will stay lighted, indicating that the edit-out point has been memorized.

Note: The IN and OUT indicators will blink alternately (1) if the same point on the tape is entered as both the edit-in and edit-out points or (2) if the edit-out point is entered before the edit-in point, prohibiting editing or previewing. It must be done right and in proper sequence.

On the Record Machine

To find the edit-in point:

1. Locate the point by turning the search dial on the right, or recorder, side of the RM 440 panel and by watching the monitor. Set the dial to 0 at this point.

2. Simultaneously depress the IN and ENTRY buttons for the recorder, to enter the edit-in point. The IN indicator lamp will go from blinking to ON.

To find the edit-out point:

1. Locate the point with the recorder search dial and its monitor, and set the dial to 0.

2. Simultaneously depress the OUT and ENTRY buttons for the recorder, to enter the edit-out point in the RM 440 memory.

After we find the edit-in and edit-out points, we start the edit by depressing the AUTO EDIT button. Assume that both tapes are midway between edit-in and edit-out points. They both run at high speed in the reverse direction and pause 5 seconds prior to (in front of) the edit-in point. Then both the playback and the recorder run forward at normal speed, the playback continues to play, and the recorder begins to record at the edit-in point. Both machines run at normal speed until the edit-out point, at which time the playback runs an additional 2 seconds and stops in PAUSE and freeze-frame. The recorder stops recording, runs 2 seconds, and reverses direction to the edit-out point, where it enters PAUSE and freeze-frame.

To monitor the edit from 5 seconds before the edit-in point to 2 seconds after the edit-out point, depress the review button and view the recorder monitor.

To stop during recording before the edit-out point, depress the END button. That point can become the new edit-out point, or a search can be made for a new edit-out point. When the dial is returned to 0 and the tape is in PAUSE, that becomes the new edit-in point.

When the edit is finished, search for and enter the next edit-in and edit-out points.

Editing videotape may be rehearsed in one of two modes:

1. *Preview mode.* After the edit-in and edit-out points have been established and entered, press the preview button. Both machines will operate exactly as in an edit, without recording. By viewing the recorder's monitor you can check the entry and exit points as well as the picture quality. The entry and exit points can be modified, or the AUTO EDIT button can be pressed to commence the actual edit

2. *Return mode.* This mode is used to check only the edit-in point. During the preview mode above, at any time after the edit-in point, press the return/jump button. The return indicator lights, the preview mode stops, and both the playback and recorder return to the edit-in point and pause in freeze-frame. If necessary at that point, repeat the preview. If the previewed edit is satisfactory, depress AUTO EDIT to make the edit.

Checking the Edit (the Recording)

The accuracy of the recording can be checked in one of two modes:

1. *Review mode.* After the edit, depress the review button. The review indicator lights, and the recorder machine operates as a playback. The playback machine does not operate.

View the recorder monitor. The recorder plays back the picture from 5 seconds before the edit-in point to 2 seconds after the edit-out point. When the review is finished, the recorder returns to the edit-out point. The review may be stopped with the PAUSE button.

2. *Jump mode.* This mode is used to quickly check the edit-out point. During the review mode above, at any time after the edit-in point, depress the jump/return button and the tape will fast-forward to the edit-out point. The jump indicator will light, the review mode is stopped, and the tape runs at high speed to a point 5 seconds before the edit-out point. From this point the tape runs at normal playback speed to 2 seconds after the edit-out point, showing a picture on the recorder monitor, and then returns to the edit-out point, where a freeze-frame is displayed. Note that the jump mode is obtainable only on recorders that have high-speed playback.

Some Editing Shortcuts

Editing time can be shortened by entering the edit-in and edit-out points during preview. Using the search dials, search for the scene starting point on the playback and recorder; then set each to 0 to pause the tape. Press the preview button, and the scene starting point just chosen for both machines will be memorized as the edit-in point for both machines. The IN lamp will light, and the preview will commence.

Search for the edit-out point and press the END button, which will cause that point to be memorized as the edit-out point for the playback. The OUT lamp will cease blinking and remain lighted. After 2 seconds, both machines will pause. Preview the tape; if necessary, change the

edit points. Finally depress the AUTO EDIT button to perform the edit.

To edit even more quickly, without entering edit-in and edit-out points, search for the scene starting point on both machines and pause both tapes at 0. Press the AUTO EDIT button, and recording will begin at that point, making it the edit-in point. At the scene endpoint, press the END button, which stops the edit at that point and makes it the edit-out point.

The edit points can be finely adjusted, one frame at a time, backward or forward by using the time counter in the trim mode. Press either the IN or OUT button, and hold it down. The frame number of the edit-in or edit-out point will be displayed on the time counter. Then press and release the TRIM + or the TRIM − button to move the tape forward or backward, respectively, one frame at a time. The counter will change, indicating the frame motion.

To do *butt* editing, or continuous editing, recall that at the end of an edit the recorder returns to the edit-out point and stops. This can be the next edit-in point for the recorder. Search for and enter the next edit-in and edit-out points for the playback. Then press the AUTO EDIT button, and the edit will be performed. Alternately, find only the edit-in point for the playback and enter it. Press the AUTO EDIT button, and the edit will start. To stop the edit, press the END button, and this point will be the edit-out point.

OFFLINE EDITING

Often a producer will want to review the tape material and plan the editing without tying up an editing room and its personnel. Tape copies, called *scratch dubs*, or *workprints*, are made of the material, so the producer can work alone with a

VTR playback and a monitor, making preliminary editing decisions. This is termed *offline editing.*

If time code is recorded on the master tape, the addition of a time code reader is used to "burn in," or superimpose, graphics of time code on the picture as the scratch dub is being made. This allows the producer to screen the tapes and make editing decisions relative to time code offline.

After the editing decisions have been made, the producer brings his materials, including a cue sheet denoting the time-code edit points, to an editing room. Then an editor begins online editing of the final tape with video and audio in optimum balance, with or without the producer being present.

TITLING OR GRAPHICS

Titling, or *graphics,* is the addition of the written word or number on the screen, in *alphanumerics,* in combination with the picture.

Graphics, in one of its many forms, are used for program titles, credits, graphs and displays, and subtitles. When graphics is combined with a video switcher's special effects, it is limited by only the imagination of the user.

Graphics are performed by using a device known as a *character generator.* Character generators accept a video composite input and provide both a preview and a program output. They employ a typewriter-style keyboard, and they differ from one another in complexity in two ways: the number of *pages* of information that they can store in memory and the number of *fonts* of print style available. Some character generators can store up to 999 pages of memory information as well as provide animated graphics.

A page of memory consists of a specific number of letters, numerals, or symbols, formatted into a specific number of lines, with *x* number of characters per line.

The character fonts are available on various generators in either or both uppercase and lowercase, varying print styles, white lettering with black edging and its reverse, color characters with differing color edging, and many other variations.

KNOX VIDEO PRODUCTS KX50 AND KX60 CHARACTER GENERATORS

The KX50 and KX60 generators each provide a single printing font in uppercase letters, numerals, and symbols, for a total of 64 characters, 26 English letters, 10 numerals, 27 typewriter symbols, and 1 space (Fig. 9–5).

The KX50 provides the larger character size, and a single page consists of 8 lines, with 32 characters per line. The KX60 provides a format of 64 characters in each line of a 16-line page. The height of the characters on the video screen is 36 scanning lines for the KX50 and 18 scanning lines for the KX60. Both machines have an internal memory of 4 pages, which is switch-selectable.

In addition to the typewriter keyboard, the front panel has three switches: Program ON/OFF controls the character video on the program channel when in the ON position. Page select allows page selection via the keyboard numerals in the SELECT position. When pages 1 through 4 have been selected, a corresponding LED indicator on the panel will light. And page-title-crawl, in the top position, provides a full-page screen display; in the middle position, it activates the title mode and places program copy in a single line on the lower third of the screen; and in the lower position, the crawl mode, the line

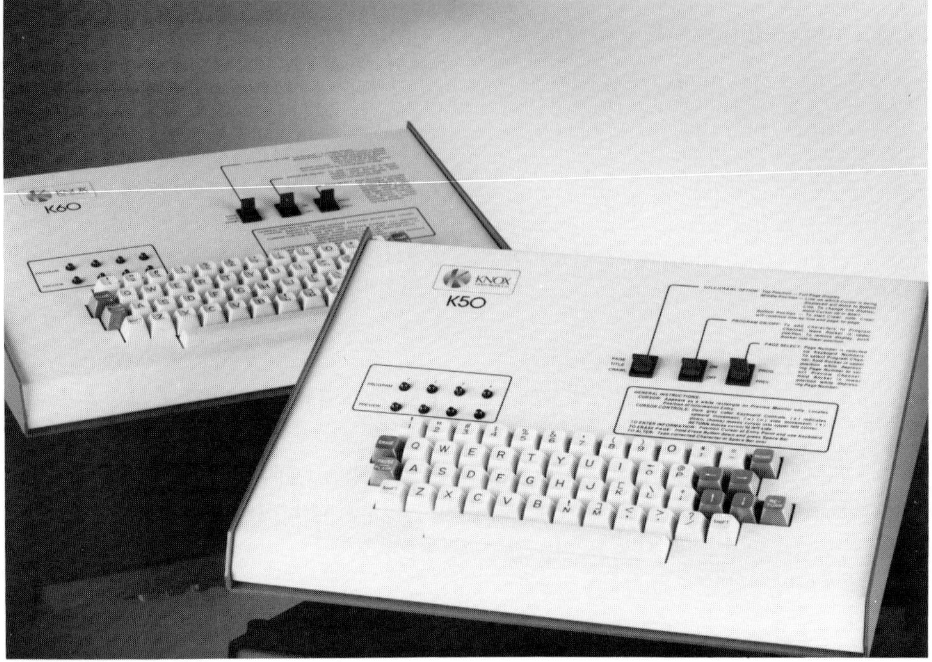

Figure 9–5 Knox KX50/KX60 Character Generators.

of copy on the lower third of the screen is set in motion, moving from right to left, and may be continuous from page to page; or there may be as many as 32 independent subtitles. There is also a page-sequencer switch on the rear of the character generator which presents the pages in order and at a rate of speed change which is adjustable and set internally.

Knox Video Products makes a more complex generator, the K128, with a standard character set and a multifont option, three-speed roll and crawl, automatic centering and double-character-size options, and an independent edit channel option. Additional options include an 8- or 16-page memory or a floppy-disk 400-page memory.

Knox Video Products also makes an accessory called *the color box*, which generates color backgrounds, color characters, and full surrounding edging for the K128 character generator.

SHINTRON 505 CHARACTER GENERATOR

The Shintron model 505 has 16-page memory with each page consisting of 8 lines, 16 characters per line. It has two character sizes, 18 or 36 scanning lines, that are height-selectable; crawl; three preselectable roll lengths; line cancellation; character blink; and foreign characters in French and Spanish.

CHYRON RGU-2
CHARACTER GENERATOR

Chyron Telesystems makes the model RGU-2 which uses literally a library of character fonts containing 45 font selections of 18 different typefaces in sizes ranging from 18 to 58 TV scan lines. Other features of the RGU-2 include proportionally spaced characters, background stripes and solids, choice of edging, see-through characters, multiple roll and crawl speeds, slow reveal, floppy-disk storage, insert and delete controls, and programmed animation.

REVIEW QUESTIONS

1. What is postproduction and why is it done?

2. What equipment is needed to do postproduction?

3. Explain why time code is sometimes used and what equipment is needed.

4. What does an editing controller control?

5. Can we directly mix the outputs of two VTRs? Explain.

6. Describe the time-base corrector. Describe jitter.

7. Why is videotape prerolled during an edit?

8. Name the two editing methods and describe the difference between them.

9. Can we rehearse an edit? Explain.

10. How can titles be added to the video?

Chapter 10

The Studio Cable TV Program

(Chapter 10)

THE TV STUDIO

Before we discuss television programs, let us look at the room where programs are produced, the television studio, and what we may expect to find there.

The room itself should be minimally 30 feet by 30 feet for cable television productions or other small productions. If it is going to include storage space for sets, flats, and drops, then a larger space is preferable. The room should be as isolated from *external* sound as possible, and its walls should be treated with sound-absorbing materials and be medium gray to dark. A cyclorama curtain is often installed, so that the walls seem endless to the camera.

The ceiling should be high, preferably 16 to 20 feet from the floor, so that a lighting grid may be hung above, and be out of the way of, the cameras and sets. A group of lighting fixtures, clamped to that grid, illuminate the set(s), and a movable ladder enables the studio electrician to hang and to adjust the lights. The individual light fixtures may be turned on separately, or all the lights may be wired to a lighting control board, where they may be controlled in any desired manner.

There should be labeled paralleled connectors located on all four walls for camera cables, mike cables, power, studio monitor outputs, and intercom units. This allows various camera and mike setups with a minimum number of feet of cable coiled on the studio floor.

THE SET

Here are a few studio set definitions: A *flat* is a large, framed canvas which may stand alone, supported by detachable supports, that is painted to look like a particular background. A *drop*, or *backdrop*, is a weighted, rolled canvas that is hung like a window shade, with a painted background. A set may include both flats and drops as well as whatever furniture and props are required to dress the stage or shooting area, so that it represents whatever scene is being created for the viewer. The studio stagehands dress the set.

The set background, particularly on a news program set, is often a drop, a drape, a curtain, or a flat with an area painted a special hue of blue. This color hue is designed to trigger *chroma keying*. The switcher with chroma keying inscrts an external signal whenever it "sees" a line of that color blue. Therefore, when the switcher sees the blue hue area over the anchor's shoulder while the chroma key mode is on, whatever external signal (film, tape, still picture) is desired by the director will be seen in place of that blue background. Extreme care must be taken to avoid use of that hue in any other part of the set, including the anchor's eyes; or else it, too, will trigger the switcher, and the insert will appear in place of the eyes.

Whenever several recurring television programs are performed in the same studio at a different times, it is common practice to store their set materials in the studio. Otherwise, a fairly large storage area adjacent to the studio is needed.

The studio will have cameras on pedestals or tripods on wheels, called *trolleys* (a combination of the words *tripod* and *dolly*), video monitors on movable stands, and audio monitor speakers. There is often a large clock which is started from the control room at the beginning of the show so the performers can time and pace their program material.

THE CONTROL ROOM

Usually adjacent to the studio, is the control room, but as we mentioned earlier, this is not necessarily so. The control room has a group of labeled video monitors (air, preview, cam 1, cam 2, telecine, etc.), audio monitor speakers, video control equipment (CCUs, waveform monitor, and vectorscope) at the video technician's position, the production switcher at the switching position, the audio console, and the program director's position.

PRODUCTIONS AND SCRIPTS

Now, what can be done in the studio? Productions. First we describe the parameters of a *major* production, such as a stage play or a documentary drama, keeping in mind that understanding how a large production is put together will help immeasurably in putting together a small one.

A production starts with an idea. Then the idea is expanded (or "fleshed out"), combined with other ideas, given a location, given characters or participants, and written into a script. When the script is accepted by the television facility, it is given to a producer. Frequently it happens the other way: and the producer brings the script to the TV station. The producer arranges for funding, a director, crew time, facilities (which may be studio time or Portapak time), and editing time. Then the producer, unless he is also directing, assumes an emeritus position and turns over the project to the director.

The director does an overlay of the script, to determine where all the participants in the production will be physically located and where each camera in the production will be at any moment, in order to capture the action of the production. This is termed *blocking out* locations, and it must be carefully conceived and indicated in the *shooting script*.

Next the director hires, or otherwise arranges to get, the *talent* and rehearses them away from the studio, in the manner in which he wants the script to be read or played. When the talent is perceived by the director to be ready, a rehearsal is called in the studio This is *facilities ("fax") time,* with a full complement of technical staff. The rehearsal is run as though the production were to be recorded (indeed, it often is), stopping only for unforeseen problems with props, cameras crossing in front of each other, mikes improperly placed, talent or production goofs, or any other difficulties that would compromise perfection in the production. Each problem is worked out and corrected, and the possible problem is noted in the shooting script, so it may be avoided during the actual take.

Finally, the production is performed again as a take, sometime later, hopefully this time perfectly, and it is recorded, to be aired when scheduled. Note that the production need not be shot from beginning to end in the order seen on the air. It may be taped in any order, or sequence, that makes sense in the use of cast, facilities, or funding and then edited during postproduction into a completed entity.

The production just cited is the most complex type of program performed in television.

Another style of production we want to look at is the *repeated format*, typified by the daily or weekly newscast. The set is always the same—a large imposing desk, behind which several people can be seated, against a backdrop which is neutral enough to permit using the switcher's special effects to key a framed tape insert or live feed which appears over the newscaster's shoulder.

The producer's function on this type of program is usually *pro forma*, and the director calls the shots. The newscast material is prepared by the news department, together with news inserts, and is given to the news *anchor* in sufficient time for him to preread and adjust the script to his reading style and be aware of the *lead-ins* and *lead-outs* of the taped inserts. At a small facility, the anchor often *is* the news department.

The program is not taped; it is done *live*, in real time. It opens with either a taped *billboard* opening or a live greeting from the anchor. As the first news item is introduced, the director cues the first tape insert, which appears over the anchor's shoulder, a keyed special effect from the switcher. The anchor introduces the

item, and the director cues a switch from the anchor camera to the insert input on the switcher. Now the tape insert occupies the full screen and continues to its conclusion, perhaps with the anchor narrating off-screen or with the anchor introducing audio from a reporter on site.

On the insert's out-cue, the director cues a switch to the anchor camera, in a *one-shot*. The anchor introduces the sportscaster, and the camera zooms wider to a *two-shot*, revealing the sportscaster sitting at the table to the left (or right) of the anchor. As the sportscaster begins to read ball scores, the camera frames him and zooms tight on a one-shot of the sportscaster alone. At the director's cue, the camera zooms back a bit so that the next cue, the taped insert of the ball game, can be keyed over the sportscaster's shoulder.

An example of a much simpler program format, and one in which the producer has the idea for the program, occurs when the producer is a townsperson who wants to pay tribute to a city or town. The idea is a short program showing montages of the town's important people, its important buildings, and its growth. The program is designed to engender excitement and build civic pride. The idea is scripted as follows:

Script

Video: Opening shot, town hall, flags waving. *Audio:* Martial music up and under the narrator: "This *is* (pause) our (pause) town." Music up, and video dissolves to mayor at office desk, smiling. *Audio:* "We welcome you to (town)." Music up, and video dissolves to town council chambers. Council is in session, with the narrator in, as music fades. Narrator talks about democracy in action, etc., as video dissolves to shopping centers, schools, ballfields, library and as audio describes the town with music under. Time of program: 5 minutes.

*A FINAL WORD ON
TELEVISION PRODUCTION*

Rather than conclude this book with a solemn discussion of television production, we discussed it throughout the text. And it is no more complex or overwhelming than any of the other topics discussed.

We prefer this approach both to discuss television production and to conclude the text:

Three five-year-olds are playing together. One says, "Let's play mommies. You be the mommy, and we'll be the children." I say, "Let's play television. You be the camera operator, you be the video recordist, and I'll be the producer-director. We will go out with some equipment and have a hell of a lot of fun."

It is my sincerest hope that it is what you will do.

REVIEW QUESTIONS

1. Why should a studio be isolated from external sound?
2. How high should the ceiling of the studio be? Why?
3. Where are sets stored?
4. What is a shooting script? Blocking out?

Chapter 11

Glossary of Television Terminology

This glossary is intended to provide a basic working vocabulary of the terms found in the industry. It does not purport to be encyclopedic and may not supply exact dictionary definitions.

Access Community-access television, channel(s) on a cable television system reserved for the use of the members of the community.

Acetate An instantaneously cut audio recording disk; the material on which it is cut.

Acoustic Pertaining to the sound environment; material for the control of sound on surfaces.

Active A circuit containing amplification.

Ambient noise Background sound.

Ampere Unit of electric current, or electron flow.

Amplifier An electronic active device that enlarges signal impulses passing through its stages.

Anechoic chamber An enclosure devoid of sound reflection.

Antenna A structure designed to either radiate or capture radio-frequency energy.

Aspect ratio The ratio of picture height to picture width; 3:4 in U.S. television.

Assemble edit A videotape edit, in which the video signal, audio signal, and control-track signal are all edited at the same time.

Attenuator A passive resistive device that presents an insertion loss to a circuit. *See* Pot, Pad.

Audio Sound, within the spectrum 20 Hz to 20 kHz.

Audition Listening to audio without broadcast transmission.

Background sound A secondary sound source which is audible behind or under the primary source.

Balance The blend of gain from two or more sources, so that they are in correct proportion to one another.

Ballistics, meter The VU meter uses a method of damping that causes the meter to read the average audio level; PPM meter ballistics cause the meter to read audio level peaks.

Bias A higher-than-audio frequency which is applied with the audio to tape during the recording process, to correct for nonlinearity of magnetization.

Bidirectional mike A mike with two lobes of sound acceptance, at 0° and 180°.

Blanking The time during which the scanning beam is turned off.

Blast filter A foam filter in or on a mike to prevent blast-in.

Blast-in A sound input of extreme excessive level.

Board, control *See* Console.

Boom A microphone stand which projects the mike horizontally, usually above the user. A baby boom is a smaller version.

Booster amplifier An amplifier of intermediate gain, used between a preamplifier and a power or line amplifier.

Bridge A short musical piece used as a sound transition; a voice-under as a news bridge.

Bridging An input impedance of 10 kΩ or more; a method of connecting electronic devices in which there is a gain loss but not a frequency-response loss.

Bus A line extension of a series of electrical points; a common point for a group of amplifier outputs; a point to which amplifiers may be switched without disturbing the impedance balance of that commonality.

Cable television Television channels that are delivered to the viewer by coaxial cable rather than being transmitted over the air; cable operator, franchisee, or the company which supplies cable television.

Cans Earphones, headphones.

Capstan When used with an associated tangential pinch roller, it maintains constant tape speed in a tape transport mechanism.

Cardioid The shape of a unidirec-

tional, heart-shaped microphone sound acceptance (polar) pattern.

Carrier A transmitting wave in the radio-frequency spectrum which transports, as its sidebands, modulation in the audio- or video-frequency spectrum.

Cartridge, phono A transducer which derives the modulation from a record groove.

Cartridge, tape A continuous closed loop of lubricated audio tape on one reel, encased in a plastic container; the record and playback equipment for that tape.

Cassette An encased, reel-to-reel audio or video tape; the record/playback system for that tape.

CATV Community antenna television.

Chain, mixing The components, including input, pot, key or assignment switch, and a preamplifier of high gain, which can feed a bus or program level channel.

Channel, audition *See* Cue.

Channel, mixing *See* Chain.

Channel, program The input, pot, and +4- to +8-db amplifier which can feed an audio program line or the input to a recorder.

Chrominance The color portion of the picture information.

Clearance Permission obtained from authors, composers, or publishers to use material for broadcast.

Clip To cut off sharply.

Coaxial cable A special concentric cable used to maintain constant terminal impedance and to keep spurious signals from intruding on the signal being carried on the coaxial cable.

Color bars A pattern of vertical stripes, each with a particular hue, that is used for checking the color performance of a television system.

Commercial message That portion of broadcast time during which the sponsor's product or service is described to the listening or viewing audience.

Compander A signal-processing device; an acronym for compressor/expander.

Composite video A combination of picture and synchronizing (timing) information.

Console An electronic device that controls gain, mixes, balances, and routes audio. Also called *control board,* or *mixer.*

Continuity Scripted material.

Control room A room adjacent to a studio in which the audio and video control functions are performed.

Crosstalk Spillover of audio or video energy from one circuit to another; spillover of magnetic energy from adjacent tape layers, known also as *print-through.*

Cue A signal to start; *cue up*—to prepare a recording for playback; *cue sheet*—a run-down sheet of cue-ups for the operator of a multi-insert program; *cue system*—the amplification and listening system on which recordings are cued up or auditioned.

db decibel, or 0.1 bel; dbm—db referenced to 1 mW of sinusoidal power in a 600-Ω line.

Dead Not live, as in dead mike, dead studio; dead pot—to playback on a closed pot.

Decibel see db.

Degauss To demagnetize, to disrupt the magnetic alignment of tape particles.

Dissolve To fade one signal out at the same time and rate that another signal is faded in.

Distortion, audio Variation from pure signal quality. Major types are intermodulation distortion (IMD), total harmonic distortion (THD), and frequency distortion.

Dolly A mobile television camera stand; to move a video camera.

Dub To make a copy of a recording.

Dynamic microphone A mike using a moving coil or a condenser diaphragm transducer.

Echo Reverberation of sound; echo chamber, an electronic or acoustic device for creating reverberation or sound delay.

Edit To remove, replace, or change portions of an audio or video tape. *See* Assemble Edit and Insert Edit.

Edit search Determination of the beginning or the end of a tape edit by looking for edit points on the tape.

EIN Equivalent Input Noise, a measure of performance in relation to noise in a circuit, where EIN measures the noise floor.

Effects The generator of special-effects portion of a production switcher.

EFP Electronic field production.

ENG Electronic news gathering.

Equalizer An electronic device, either active or passive, which boosts or cuts a frequency or band of frequencies.

Fade To lower or increase the gain of an amplifier by using a pot or vertical slide attenuator; fader—the control used to vary gain.

FCC Federal Communications Commission, the U.S. regulatory agency which licenses and controls broadcasters.

Feed   To transmit electronically from one component or circuit to another; *feed reel*—the tape reel which dispenses the tape.

Feedback   Feedback loop, a closed loop of audio energy in a run-away state, causing an ever louder oscillatory howl. It is called *howl round* in the United Kingdom.

Field   *See* Frame.

Filter   A passive device to cut or boost frequencies.

Flat-topping   Cutting the tops of audio waves, causing distortion.

Floating mults   Patch panel jacks which are connected to each other but float (unconnected to a working circuit) unless used.

Fluff   A speech error.

Flutter   A speed irregularity in a playback device.

Footcandle   A measurement of light intensity. *See* Lux.

Frame   A television picture exposure. There are 30 frames per second in the NTSC system; each frame is divided into two fields.

Frequency   The number of vibrations or cycles in a given time period, as cycles per second, measured in hertz; *frequency response*—the degree to which audio equipment maintains its designed spectrum coverage without distortion.

Gain   Audio or video energy volume; gain figure, the specific amplification inherent in an amplifier.

Graphics   Titles or other printed alphanumeric characters placed within the video picture by a character generator.

Grease pencil   A soft, waxy marker used in audio tape editing.

Groove   The depression cut in a record by the stylus. The left and right channels of stereo music are cut in the left and right sides, respectively, of a record groove.

Ground   The electric potential reference of zero; to connect components together and then to earth for a commonality of potential.

Headroom   The amount of space between the top of a modulation cycle and the point where clipping of that top occurs.

Head, tape   The transducer in a tape recorder; *head gap*—the space between the pole pieces of a tape transducer.

Helical scan   A television tape recording system in which the modulation is written diagonally on the tape from a rotating-head cylinder.

Hertz   (Hz) cycles per second.

Impedance Z A measurement of electric resistance to current flow in an alternating-current circuit.

Input   That part of a circuit or component into which energy is fed; the front end of a circuit.

Insert   A recorded piece of a program.

Insert edit   The video signal and audio signal can be edited independently or at the same time. The control track is already on the tape and is the editing reference.

Jack   A female connector with a male name; *jack panel* (or *bay*)—a field of female connectors. *See* Patch Panel.

Kelvin—degrees   The unit of color temperature.

Key   A switch used to interconnect components on a control console; *key in*—to bring in modulation with a key; *key station*—a network-originating station.

Kilohertz (kHz)   1000 Hz.

Label   The center portion of a record that contains the hole and information such as the artist, the selections, the selection time, and the recording company.

Leader tape Nonmagnetic plastic or paper tape used to protect the ends of, or separate the segments of, recorded tape.

LED Light-emitting diode, a small illuminator used to indicate on/off conditions.

Lens, zoom A multielement lens system with variable focal length.

Level The amount of audio energy, as read on a VU meter.

Light Refraction, the bending of light rays as they pass from one transparent medium to another; fixtures, instruments, lighting devices; measurement, in footcandles or lux, with an exposure meter.

Limiter A circuit that protects against signal overload.

Line A two-wire circuit.

Linearity A relationship between two quantities in which a change in one quantity is directly proportional to a change in the other.

Live Not recorded; produced in real time.

Log A document relating to programming or technical facts as they occur at a broadcast station, which is kept as a requirement of the FCC.

Loop A wire circuit; a repetitive closed continuum of tape for reverberation purposes; a video signal feed-through. *See* Feedback Loop.

Loudspeaker A transducer that changes electric energy into mechanical (sound) energy.

Lumen A measure of light output. Onc lumen falling on one square foot of surface, at a distance of one foot, produces one footcandle of light intensity.

Luminance Black-and-white picture information.

Lux A measure of light intensity, where 1 footcandle of light equals 10 lux.

Master control A central program switching point.

Master pot The gain control on the output of a program amplifier.

Matrix A group of switches for routing program information.

Microprocessor An electronic chip (integrated circuit) for processing data.

Mike Microphone; a transducer that changes mechanical (sound) energy into electric energy; *miking*—using microphones to pick up sound.

Mixer A gain control; a group of gain controls feeding a bus; a consolette used on remotes.

Modulation Sound energy carried by another medium.

Monaural A single program channel, or track.

Monitor To listen to program.

Mult *See* Floating Mults; a unity-gain amplifier with multiple outputs.

Multiplexing Using the two sidebands of a modulated carrier wave to carry separate program information.

Muting relay A relay that is operated by a mike key and cuts studio speakers when a mike is live.

Network A company that controls the program output of a group of broadcast stations.

Noise Random sounds which are not harmonically related and which may occur in any portion of the audio spectrum. *See* Snow.

NTSC National Television Standards Committee, which sets the U.S. standards for color television.

Off-mike Not within the pickup pattern of a mike.

Ohm The unit of electric resistance, or impedance.

Omnidirectional In all directions.

On-mike On axis to the pickup pattern.

Optics system Television camera, a set of prisms or mirrors which divide the light input from the lens to the three pickup tubes; a light-beam splitter.

Output The portion of a circuit from which energy flows.

Pad A resistive insertion loss in a circuit.

Panpot A gain control which is used to fade audio from one circuit to another.

Passive circuit A circuit having no amplification.

Patching A system of externally connecting or disconnecting circuits or components by using patch cords and patch bays.

Pattern, polar A polar graph of the ability of a mike to react to sound waves as they impinge on its transducer either on or off axis.

Peak A rise from the flat portion of a curve; a high reading on a VU meter.

Phase Mikes are *in phase* when their outputs are additive and *out of phase* when their outputs are subtractive.

Pickup A phonographic transducer; sound received by mikes; a feed from a mult or PA system.

Pink noise White noise fed through a special audio filter which inverts the frequency characteristics of the white noise and results in a test signal of uniform level. *See* White Noise.

PFL Prefader listening.

Platter A phonograph record.

Playback The device used to feed audio or video from a recording.

Popping An explosive sound caused by stress in the pronounciation of the letters B, P, and T.

Pot Potentiometer; gain control; alternately called a fader.

PPM Peak programme meter, United Kingdom; see Volume Unit Meter.

Preamplifier A high-gain amplifier.

Presence The quality of being on mike at the correct individual distance from the mike.

Pressing A manufactured copy of a disk recording.

Pressure mike *See* Dynamic Mike.

Print-through *See* Crosstalk.

Program A broadcast presentation with a beginning and an end; program channel; *see* Channel.

Puller Tape transport mechanism.

Receiver A device that demodulates audio or video information from a carrier wave.

Record To make a recording.

Relay An electrically operated switch, either mechanical or transistor.

Remix To mix down multichannel recordings to two or four channels.

Remote A broadcast or recording made at other than the station's studios.

Residual magnetism The magnetic flux that remains in a substance after the magnetizing force has been removed.

Response curve A graph indicating a mike's gain in decibels versus frequency in hertz; *see* Frequency Response.

Reverberation Sound reflection; echo.

Ribbon mike *See* Bidirectional Mike.

Ride gain To control the gain of audio.

Roll-off A diminution, or fading, at one end of a curve.

Scrim A fiber or plastic light diffuser.

Segue Pronounced "seg-way," a continuous playing of two or more music records with no break or announcement in between.

Servo A motorized control device.

Setup, camera All the controls adjustments that are necessary at periodic intervals to keep a television camera in optimum operating condition.

Shaped response A design factor in microphone frequency-response curves resulting in a roll-off or peak.

Shoot Use of a video camera to produce program material.

Sibilance The hissing sound made in the overstatement of the letter S.

Signal-to-noise (S/N) ratio A qualitative statement about the performance of an electronic device based on the ratio of the two stated factors. It may also be stated in reverse, as a noise-to-signal ratio.

Signal processing Alteration of audio signal to remove unwanted (noise) portions of the spectrum or to add boost to specific frequency(s).

Sine wave An audio wave whose shape can be expressed as the sine of a linear function of time, space, or both; a tone signal, a signal with equal positive and negative values, above and below a fixed baseline.

Snow Video noise.

Solo Alone; an audio monitor feature in which one signal out of a group may be isolated for listening without removing that signal from the group.

Sound A disturbance of air particles causing audible vibration of those particles.

Spillover A VU reading of over 0 VU and into the plus, or red, portion of the scale.

Sponsor One who pays a fee to advertise a product or service on a broadcast.

Spot Spot announcement; a short advertisement.

Stake-out Waiting on an ENG assignment for a statement or interview following a closed-door meeting.

Standing wave A reflection of energy in opposition to the normal flow, in a circuit, from a point where an impedance mismatch occurs.

Stereo Stereophonic; two complementary channels of recorded or broadcast information.

Studio A room designed to be used for recording or broadcasting.

Stylus The shaped needle with industrial-diamond tip which either cuts the record groove or transfers the undulations of the groove to the playback transducer.

Switcher The device used in production to mix and switch between video inputs.

Sync generator The device that supplies synchronizing (timing) signals for the exacting timing needed in a video system.

Talent The persons on mike or on camera in broadcasting.

Talk-back An intercom system between the control room and studio.

Tape A plastic ribbon of specific width that is coated with a magnetic oxide and used as a recording medium.

Target   The photosensitive metallic grid of the television pickup tube.

Telco   The telephone company.

Time-base corrector   A computer that exactly retimes the horizontal picture lines emanating from a video tape recorder.

Track   The magnetic signal that is written on recording tape.

Transcription   A record that is manufactured to be used solely for broadcast.

Transducer   A device that changes one form of energy to another form.

Turntable   A playback machine for records and transcriptions.

Unidirectional   One-directional; *see* Cardioid.

Velocity mike   *See* Bidirectional Mike.

Viewfinder   A small television monitor mounted on the top of a television camera to permit the camera operator to see picture.

Volt   The unit of electric pressure.

Volume   A quantitative amount of sound; *volume unit*—dbm of complex sound waves; *VU meter*—an instrument which measures sound volume.

White noise   Noise which is uniform in level over a frequency band; all noise frequencies perceivable to the human ear, heard together.

Wow   Sound that is off pitch, caused by an off-speed playback.

Zoom   *See* Lens.

Index